S0-BQI-349

Evolutionary Conservation of Developmental Mechanisms

The Fiftieth Annual Symposium of
the Society for Developmental Biology
Marquette University, June 20-23, 1991

Executive Committee
1990-1991

1991-1992

Business Manager
Holly Schauer
P.O. Box 40741
Washington, DC 20016

Evolutionary Conservation of Developmental Mechanisms

Editor

Allan C. Spradling

Howard Hughes Medical Institute
Carnegie Institution of Washington
Baltimore, Maryland

A JOHN WILEY & SONS, INC., PUBLICATION
New York • Chichester • Brisbane • Toronto • Singapore

Address All Inquiries to the Publisher
Wiley-Liss, Inc., 605 Third Avenue, New York, NY 10158-0012

Printed in the United States of America.

Library of Congress Cataloging-in-Publication Data

Society for Developmental Biology. Symposium (50th : 1991 :
Milwaukee, Wis.)
Evolutionary conservation of developmental mechanisms / the
Fiftieth Annual Symposium of Society for Developmental Biology,
Milwaukee, Wisconsin, June 20-23, 1991 ; editor, Allan C. Spradling.
p. cm.
Includes bibliographical references and index.
ISBN 0-471-58843-1
1. Genetic regulation--Congresses. 2. Developmental cytology-
-Congresses. I. Spradling, Allan C. II. Title.
[DNLM: 1.Developmental Biology--congresses. 2. Evolution-
-congresses. 3. Gene Expression Regulation--congresses. QH 453
S679e 1991]
QH458.S63 1992
575.1--dc20
DNLM/DLC
for Library of Congress 92-18116
CIP

The text of this book is printed on acid-free paper.

Contents

Contributors

J.A. Banks, Department of Botany and Plant Pathology, Purdue University, West Lafayette, IN 47907 **[9]**

Richard W. Beeman, USDA, ARS, U.S. Grain Marketing Research Laboratory, Manhattan, KS 66502 **[71]**

Janet Braam, Department of Biochemistry and Cell Biology, Rice University, Houston, TX 77251 **[185]**

Donald D. Brown, Department of Embryology, Carnegie Institution of Washington, Baltimore, MD 21210 **[1]**

Susan J. Brown, Division of Biology, Kansas State University, Manhattan, KS 66506 **[71]**

Helen M. Chamberlin, Howard Hughes Medical Institute, Division of Biology-California Institute of Technology, Pasadena, CA 91125 **[141]**

A.B. Cubitt, Department of Biology, Center for Molecular Genetics, University of California, San Diego, La Jolla, CA 92093-0634 **[159]**

Rob Denell, Division of Biology, Kansas State University, Manhattan, KS 66506 **[71]**

R.K. Esch, Department of Biology, Center for Molecular Genetics, University of California, San Diego, La Jolla, CA 92093-0634 **[159]**

N.V. Fedoroff, Department of Embryology, Carnegie Institution of Washington, Baltimore, MD 21210 **[9]**

R.A. Firtel, Department of Biology, Center for Molecular Genetics, University of California, San Diego, La Jolla, CA 92093-0634 **[159]**

C. Gaskins, Department of Biology, Center for Molecular Genetics, University of California, San Diego, La Jolla, CA 92093-0634 **[159]**

Robert Glaser, Howard Hughes Medical Institute, Carnegie Institution of Washington, Baltimore, MD 21210 **[39]**

J.H. Hadwiger, Department of Biology, Center for Molecular Genetics, University of California, San Diego, La Jolla, CA 92093-0634 **[159]**

Sarah Hake, USDA-ARS Plant Gene Expression Center, Albany, CA 94710 and Department of Plant Biology, University of California, Berkeley, CA 94720 **[111]**

Russell J. Hill, Howard Hughes Medical Institute, Division of Biology-California Institute of Technology, Pasadena, CA 91125 **[141]**

P. Howard, Department of Biology, Center for Molecular Genetics, University of California, San Diego, La Jolla, CA 92093-0634 **[159]**

Elizabeth A. Hurley, Howard Hughes Medical Institute, Departments of Internal Medicine and Microbiology/Immunology, University of Michigan Medical Center, Ann Arbor, MI 48109 **[21]**

Gary Karpen, Howard Hughes Medical Institute, Carnegie Institution of Washing-

The numbers in brackets are the opening page numbers of the contributors' articles.

ton, Baltimore, MD 21210, present address: MBVL, The Salk Institute, La Jolla, CA 92037 **[39]**

Randy Kerstetter, Department of Plant Biology, University of California, Berkeley, CA 94720 **[111]**

Richard Kostriken, Department of Molecular and Cell Biology, University of California, Berkeley, CA 94720 **[125]**

Brenda Lowe, USDA-ARS Plant Gene Expression Center, Albany, CA 94710 **[111]**

S.K.O. Mann, Department of Biology, Center for Molecular Genetics, University of California, San Diego, La Jolla, CA 92093-0634 **[159]**

P. Masson, Laboratory of Genetics, University of Wisconsin, Madison, WI 35706 **[9]**

K. Okaichi, Department of Biology, Center for Molecular Genetics, University of California, San Diego, La Jolla, CA 92093-0634 **[159]**

Nipam H. Patel, Department of Embryology, Carnegie Institution of Washington, Baltimore, MD 21210 **[85]**

J.A. Powell, Department of Biology, Center for Molecular Genetics, University of California, San Diego, La Jolla, CA 92093-0634 **[159]**

Andrew Ransick, Division of Biology, California Institute of Technology, Pasadena, CA 91125 **[55]**

G.R. Schniztler, Department of Biology, Center for Molecular Genetics, University of California, San Diego, La Jolla, CA 92093-0634 **[159]**

Allan C. Spradling, Howard Hughes Medical Institute, Carnegie Institution of Washington, Baltimore, MD 21210 **[39]**

Charles A. Staben, School of Biological Sciences, University of Kentucky, Lexington, KY 40506 **[199]**

Paul W. Sternberg, Howard Hughes Medical Institute, Division of Biology-California Institute of Technology, Pasadena, CA 91125 **[141]**

Jeffrey J. Stuart, USDA, ARS, U.S. Grain Marketing Research Laboratory, Manhattan, KS 66502, present address: Department of Entomology, Purdue University, West Lafayette, IN 47907 **[71]**

Craig B. Thompson, Howard Hughes Medical Institute and Departments of Internal Medicine and Microbiology/ Immunology, University of Michigan Medical Center, Ann Arbor, MI 48109 **[21]**

Larry W. Tjoelker, Howard Hughes Medical Institute and Departments of Internal Medicine and Microbiology/ Immunology, University of Michigan Medical Center, Ann Arbor, MI 48109 **[21]**

Bruce Veit, Department of Plant Biology, University of California, Berkeley, CA 94720 **[111]**

Erik Vollbrecht, USDA-ARS Plant Gene Expression Center, Albany, CA 94710 **[111]**

Cathy J. Wedeen, Department of Molecular and Cell Biology, University of California, Berkeley, CA 94720 **[125]**

David A. Weisblat, Department of Molecular and Cell Biology, University of California, Berkeley, CA 94720 **[125]**

Ping Zhang, Howard Hughes Medical Institute, Carnegie Institution of Washington, Baltimore, MD 21210 **[39]**

Preface

Two valuable and complementary modes of thought about development and evolution have colored research in developmental biology. On the one hand, organisms must be recognized as almost infinitely particular. The different species within a genus, and even individual organisms, have peculiarities that can be understood only in terms of their current lifestyles, their evolutionary history, and perhaps even the vagaries of chance. Functions such as flight, and structures such as gills, have evolved independently in different groups from different precursors. However, presently we are learning that there is more underlying unity in biology than many have supposed. The basic mechanisms of cell physiology, and the genes that underlie them, changed relatively little as organisms evolved, and many of the mechanisms driving the evolution of multicellular eukaryotes now appear to have been common, despite their diverse results. The 50th Annual Symposium of the Society for Developmental Biology, held at Marquette University on June 20-23, 1991, attempted to focus on unifying themes in the study of developmental biology. This concentration promised not only to be intellectually productive, but also to bring the diverse research systems and interests of society members together at our 50th meeting, held during what must certainly rank as the most exciting year in the history of developmental biology.

Studies of development and evolution have been isolated by artificial disciplinary boundaries since the early part of this century. Indeed, even in 1991, the program of this symposium was criticized by some as having little relevance to evolution. In fact, by concentrating on conserved developmental mechanisms, we were likely addressing one of the most interesting and mysterious problems in the history of life on earth. Why, approximately 600 million years ago, did a profusion of diverse multicellular eukaryotic organisms appear so suddenly? This "Cambrian radiation" was likely generated when the basic mechanisms required for multicellular development as we know it were finally perfected. At the symposium we focused on several such mechanisms: the generation of a distinct germline and soma, the role of certain highly conserved transcription factors in embryonic patterning, and the transmission and interpretation of information between cells. We also looked for underlying mechanisms in some areas, such as sex determination, that appear to have changed frequently throughout the course of evolution.

The symposium would not have been possible without the efforts of many people. Gail Waring and Kathy Karrer worked tirelessly with Sandra Priegel and others from Marquette University to ensure that all local arrangements were conducive to our purpose. Holly Schauer continued, as she has for many years, to assist and support the planning of the annual meetings. We thank Donald Brown, who shared his insights on the intellectual evolution of developmental biology and related fields on the occasion of this 50th symposium. We also thank all those authors who, given the many demands on the time of researchers today, submitted articles for this volume. Finally, we are especially grateful to the National Science Foundation for its consistent support of the society's annual meeting.

Allan C. Spradling

Young Investigators Awards, 1991

First Place

Linda N. Keyes
Prnceton University
Princeton, NJ 08554

Sex-Specific Activation of the *Sex-Lethal* Early Promoter in Drosophila Embryogenesis

Second Place

Billie J. Swalla
University of California, Davis
Bodega Marine Laboratory
Bodega Bay, CA 94923

Isolation of cDNA Clones Encoding Maternal mRNAs Specific for Urodele Ascidian Development Using Subtractive Hybridization

Laurie Jackson-Grusby
Genetics Department
Harvard Medical School
Boston, MA 02115

Mouse Limb Deformity: Molecular and Genetic Analysis of a Gene Involved in Limb Pattern Formation

Abstract of the First Place Young Investigator Award

Sex-Specific Activation of the *Sex-Lethal* Early Promoter in Drosophila Embryogenesis

Linda N. Keyes, T.W. Cline, and P. Schedl
Princeton University, Princeton, NJ (L.N.K., P.S.);
University of California, Berkeley, CA (T.W.C.)

In Drosophila, the choice between male and female development is made by the switch gene, *Sxl*, in response to the X:A ratio. *Sxl* is on in females and off in males; throughout most of development this choice is maintained by sex-specific alternative splicing. We show that *Sxl* is intially activated in female embryos by transcription from a sex- and stage-specific promoter that responds to the X:A ratio. The protein products of these early transcripts can initiate the female-specific splicing of transcripts expressed from the late, sex-non-specific promoter; this in turn allows for the establishment of a stable pathway commitment through post-transcriptional autoregulation. We are using reporter gene fusion constructs to identify cis-acting regulatory sequences and to analyse the interactions between this promoter and the genes that regulate it.

Evolutionary Conservation of Developmental Mechanisms, pages 1–8

1. Developmental Biology Has Come of Age

Donald D. Brown

Department of Embryology, Carnegie Institution of
Washington, Baltimore, Maryland 21210

"Genetics and the study of development have converged during the intervening quarter-century, together with the study of molecules and macromolecules; of proteins, enzymes, nucleoproteins and others; of cells and organelles; of metabolic pathways and immune reactions; of microbes and protozoans and fungi. Their convergence has transformed biology." This was Jane Oppenheimer's introductory statement to her history of the Society's first 25 years (Oppenheimer, 1966). If she believed this 25 years ago, what would she say today? To understand the changes in our field, one has to consider the past century and the nature of what we call developmental biology.

Developmental biology is above all a set of questions. It is not a discipline. The fundamental questions were asked a century ago. Read Boveri's (1902) magnificent paper "On Multipolar Mitosis as a Means of Analysis of the Cell Nucleus" to find posed there just about every question in the field. For anyone in need of a project, E. B. Wilson's textbook "The Cell in Development and Heredity" (1924) remains a useful summary of development phenomena that need explanation.

The great developmental biologists at the turn of the century, such as Boveri and Morgan, were also geneticists. They considered developmental problems in genetic terms. The era of developmental genetics was replaced by that of experimental embryology, and from the 1930s into the 1960s developmental biology was isolated, with its own theories, methods, and even experimental animals. Experimental embryologists treated the field as a discipline rather than as a set of novel questions, and concentrated on embryonic induction, fields, gradients, and morphogenetic movement. They mapped the lineage of tissues and organs, but their methods were inadequate to lead them much beyond a description of these events. Experimental embryologists were oblivious to the burgeoning disciplines of biochemistry, genetics, and immunology. Of course, these other fields were equally insular and segregated at that time, but they had an advantage. As luck would have it, at least some of the questions that they were asking could be answered without leaving the confines of their own disciplines. Even by 1966, when the

Society celebrated its 25th anniversary, few developmental biologists used the tools of the great disciplines of biology. We were a third-world scientific community, blissfully doing our thing in isolation.

The insularity of disciplines before the 1970s is reflected by popular slogans—the condensed wisdom of the elders. The biochemists' doctrine was "never waste a clean thought on a dirty enzyme." In the early 1970s, as an editor for the Journal of Biological Chemistry who was considered to be a developmental biologist, I received two kinds of papers to review. The first had to do with purification of an enzyme from an embryonic source. The second was any paper dealing with chromatin. In those days, scientists studying chromatin considered themselves biochemists, and biochemists, when forced to categorize a paper on chromatin, thought it had something to do with developmental biology. After all, you should never waste a clean thought on a dirty enzyme. Developmental biology was viewed as hopelessly complicated and, therefore, to be avoided by a traditional biochemist. Biochemistry, when applied to development, measured the increase or decrease of an enzymatic activity as development proceeded. The sophisticated biochemist distinguished between a genuine change in the amount of a protein and the change simply in enzyme activity. It was the intellectual equivalent of a Northern blot versus a "run-on" assay.

The embryology elders held the opposite point of view. I remember hearing a famous embryologist say that over the years he had developed a working relationship with the chicken embryo. The embryo promised to occasionally divulge a secret if he, in turn, promised never to homogenize it. In that era, a bona fide embryologist scorned the reductionist approach of a biochemist. If it couldn't be studied in an intact animal, it would not yield answers of interest to an embryologist.

Another famous embryologist summarized the many years he had labored on the wonderful problems of developmental biology and modestly conceded that his life's work might have yielded some new information, but essentially the fundamental questions remained mysteries. He admired the youthful enthusiasm of the audience, but, he said, we must remember that these problems are great and complex and would be there long after we had reached his venerable age. The science of developmental biology was raised to a religious experience. Sharing it would reward us all, while leaving the problems untouched for future generations rather like a renewable resource. I suppose that this is the natural psychology for a scientist who has spent a lifetime in a field studying an intractable problem. The problem becomes a friend to be respected rather than an adversary worthy of a no-holds-barred approach. If he couldn't solve it, neither would anyone else.

This warning greeted me soon after I began my research as a young whippersnapper in developmental biology in the early 1960s, so it was refreshing to hear an iconoclastic molecular biologist like Sol Spiegelman who represented the opposite point of view. He had participated in the spectacular discoveries of prokaryotic molecular biology of the 1950s and 1960s, and he was uninhibited by, even scornful of, this adoration and submission to the overwhelming complexity of developmental biology. His successes made him impatient, irreverent, and arrogant as he decided to take on developmental biology in the style of the new breed of molecular biologists. One of the first things he did was call for a national crash program devoted to understanding the development of the sea urchin. All developmental biologists should give up their pet organism and study sea urchins. Remember that success in the early days of molecular biology could be traced to several gurus, Luria and Delbruck in particular, who convinced a generation of physicists and bright disciples to study bacteriophages and to stimulate each other through summer meetings at Cold Spring Harbor, where they exchanged the knowledge learned during the previous year. Spiegelman picked the sea urchin for this mini-Manhattan project because of its familiar technical advantages—large amounts of material, synchronous fertilization, and what seemed to him to be simplicity. I doubt that he knew very much about the previous 75 years of research in embryology on sea urchins. For Spiegelman, development was just phage research all over again but a bit more complex. When successful scientists transfer fields, the result can be refreshing, even exhilarating, leading to great change. A little naivete goes a long way. However, we can all think of striking failures. What Spieglman and others made clear was that the time was ripe for cross-fertilization, for the application of the new methods of biochemistry and molecular biology to developmental biology. The field was underpopulated; the questions were exciting; the scene was set for a revolution. Spiegelman himself soon moved on to cancer research where he played the same role.

Meanwhile, where was genetics? You recall the latest scientific aphorism. I heard it relatively recently at a Cold Spring Harbor Symposium where a geneticist made the following comment: "Biochemistry without genetics is doo-doo." (I'm not sure of the spelling, but we all know what it is.) My immediate reaction was to wonder where genetics would be today without biochemistry. Some of us are old enough to remember the "one band or one chromomere, one gene" hypothesis derived from saturation mutation analysis of selected regions of *Drosophila* chromosomes. This was as far as traditional genetics went, and would have ever gone without the use of biochemistry.

One of the best articles in the Society's 25th Symposium is called "Dynamics of Determination" by Ernest Hadorn (1966), the great Swiss developmental geneticist. After the era of Boveri and Morgan, eukaryotic genetics walled itself off from developmental biology in much the same way that embryology did from genetics. Hadron was the exception. He single-handedly kept up the tradition of Boveri and Morgan by applying genetics to *Drosophila* development. Hadron studied the fate of imaginal discs, combining genetic and surgical tools with great creativity. His analysis of "transdetermination" remains one of the most dramatic examples of stability of the determined state of cells, a phenomenon yet to be explained. The other major player was E. B. Lewis, who from the mid-forties, characterized the bithorax complex to the most detailed extent possible using traditional genetics. Hadron's students and their descendants now populate developmental genetics. Ed Lewis had few disciples but produced a complete genetic kit, with all mutants and details, that was ready for the first team of molecular biologists clever enough to understand its importance. The historical point of interest here is that Hadorn and Lewis were just about the entire effort in developmental genetics in the interval between Boveri and Morgan, and the recombinant DNA era. As an example of the segregation of *Drosophila* genetics from the other life sciences, I am not aware of a single *Drosophila* mutant discovered and characterized before the recombinant DNA era that encoded a known enzyme or protein. Contrast that with the bacterial genetics that flourished in the 1950s and 1960s by identifying auxotrophic mutants of genes encoding specific enzymes along metabolic pathways.

Twenty years ago, developmental biologists studied mainly chickens, frogs, and sea urchins. Immunologists studied rabbits, guinea pigs, and humans. Eukaryotic geneticists studied yeast, neurospora, maize, and fruit flies. Neurobiologists studied mammals, and squid axons. Biochemists studied pigeon liver, calf thymus, and *Escherichia coli*. Plant biologists were completely isolated from animal biologists. These fields had their own university departments, they went to their own meetings; segregation was all but complete.

What Jane Oppenheimer decided had come to pass 25 years ago has in fact happened in the past 10 years. The boundaries and barriers of fields and disciplines have disappeared, and we can now just call ourselves biologists and use any method that works to solve our particular question. This revolution occurred for a very mundane reason—the advance of methods. What now characterizes us as developmental biologists, neurobiologists, immunologists, or whatever, is simply the question we ask, because we are all likely to use the same methods.

This revolution in methods has especially liberated developmental biologists, who ask certain questions but rely on the methods generated by other fields and disciplines. The freedom allows us to concentrate on the unanswered questions posed by Boveri and others. Not surprisingly, the titles of the sessions at this 50th Symposium are not so different from those presented at the 25th Symposium that was entitled "Major Problems in Developmental Biology," but what a difference in the scientific methods applied to these problems.

Since we are no longer a third-world science, what then is our unfinished business? Some of you will remember Gunther Stent's controversial editorial in Science about 20 years ago following Jacob and Monod's elucidation of gene control in bacteria (Stent, 1968). He proclaimed that certain fields, developmental biology for one, had entered an "academic phase." There only needed to be some dotting of the i's and crossing of the t's. I will rephrase his point: if you can imagine the answer to a question in molecular or genetic terms, even if you are wrong, some of the mystery ("the romance") of the question has been removed. Getting it right is a matter of details. Stent considered the working of the nervous system as one of the true frontiers left for biologists.

In this era of transcriptional control, or as we used to call it 25 years ago "differential gene action," it may well be that the details of how cro and lambda repressor interact with the lambda phage operator represents a model for eukaryotic gene control (Ptashne, 1986). After all, specific proteins interact with specific DNA sequences in eukaryotes as they do in prokaryotes. I submit that the challenge of understanding many well-known developmental processes is worth the effort of elaborating the details of each individual regulatory event. It is trivial these days to isolate yet another transcription factor, and even to determine the specific DNA sequence to which it binds. It is not trivial to sort out the interaction between the multiple proteins and DNA sequences required for tissue specificity, or for some complex but integrated developmental event. The remarkable combinatorial involvement of common and tissue-specific factors, sometimes activating, sometimes repressing gene activity, all in the presence of ubiquitous chromatin, will need to be painstakingly elaborated for each of the tissues and physiological events that have intrigued us over the years. This challenge exceeds the skills of most one-dimensional molecular biologists. There are, however, classic examples of how to go about dissecting and reassembling complex biochemical systems, such as the successful unraveling of the translational apparatus and DNA replication. We will need superb biochemical skills, followed by an ultimate biophysical resolution. Biologists yearn for new methods that can substitute for crystallography. Even if the general rules of protein-DNA and protein-protein interaction have already been discovered, each step of development and organogenesis deserves its own detailed analysis. There will be plenty of surprises along the way.

Understanding the control of gene expression will answer many of developmental biology's most venerable questions. The identification of determinants has become a growth industry. The first determinant was discovered in 1912 (Gudernatsch, 1912) and proved to be thyroid hormone, which controls amphibian metamorphosis. The first RNA determinant reported is encoded by the bicoid gene in *Drosophila* (Berleth et al., 1988). Localization of bicoid mRNA in the *Drosophila* egg establishes an anterior axis. I presume there will be proteins that are stored as determinants. How determinants work will be explained, at least in part, in terms of gene expression.

The molecular basis of determination or the well-known stability of the determined state was well documented by Hadorn. Is this due to reverberating circuits which maintain high levels of key gene products whose distribution to daughter cells determine what genes two daughter genes will express (Nasmyth et al., 1987)? Or is it due to stability of transcription complexes that surround genes (Brown, 1984)?

Chromatin research is being dragged kicking and screaming into the 1990s. There have been two separate cultures in research on gene expression—those studying transcription factors as though chromatin didn't exist and those who did the opposite. Rest assured that they will have something to do with each other.

Along with molecular explanations of determination, gene expression research will help to elucidate the concepts of fields, gradients, and patterns—the phenomena that so intrigued the experimental embryologists. Rapidly dividing cells synthesize one set of proteins; stationary cells synthesize another. So many different cell types do this that one general explanation is needed. Why are cells usually incapable of expressing multiple phenotypes at the same time? If one set of specialized genes in a cell is activated, why does a preexisting set need to turn off. What do introns really do? The answer cannot be that they merely separate domains of proteins that can move about as substrates for evolution (Gilbert, 1978). This is one question that was not posed by Boveri, but I suspect its answer will have developmental significance. I like the idea that introns play a crucial, as yet unknown, role in the control of transcription (Brinster et al., 1988) and are not just a repository for enhancers, or a means to provide for alternative splicing. How do stem cell lineages work, giving rise to two different daughter cells?

Pattern formation and morphogenetic movement are in good hands since they can be studied in genetic organisms. However, organogenesis of interesting organs, especially those of vertebrates, cannot be studied in traditional genetic organisms. Will the genetics of Zebra fish or mouse ever duplicate the precision and detail that genetics of *Drosophila* and *Caenorhabditis elegans* provide? I doubt it. We need new methods that circumvent traditional genctics if we want to study many complex developmental phenomena that simply do not exist in these genetic organisms. IIow about regeneration, the immune system, organogenesis of kidney, liver, lung, bone, skin, and intestine, or the vertebrate brain? Is *Arabidopsis* a model for all of the major systems that interest us in plants? If so, then plant research is in much better shape than animal research.

The way to study a complex event like organogenesis has been to carry out a genetic screen. The screen identifies all (or most) of the genes involved in some important complex event, provides tags for simple gene isolation, and by the nature of the screen associates a function with each gene. In our laboratory we have devised a "gene expression screen" which can identify all genes whose expression is changed as a result of development, hormonal treatment, etc. It is a subtractive library method that compares two RNA mixtures (Wang and Brown, 1991). Its novelty is that it is designed to isolate all up- or down-regulated genes. The method can be applied to any event in any organism. It is inferior to a genetic screen in that it requires a separate assay to establish function.

We must develop new methods to assay gene function in nongenetic organisms. Currently, we rely on transfection or transformation methods which have variable success depending upon the system. A reliable method that delivers genes to the nuclei of living, developing embryos is essential to complete the process of circumventing traditional genetics. There is high hope for genetic knockout and replacement in mice. However, this method is a long way from replacing the traditional genetic screen. Currently, the most efficient way to determine function is a computer search for a related DNA sequence in the data base.

Finally, it is time that we begin to pay attention to related medical fields. Developmental biology is relevant to some of the most profound problems that society faces. Our closest scientific relative in the medical area is called teratology, the study of birth defects—a backwater if there ever was one. Another related field is reproductive biology, which is concerned mainly with gametogenesis, problems of fertility, and contraception.

These are incredibly important research areas because the single most important thing that we could do as a nation is not control nuclear energy, clean up the environment, nor control weapons. It is to provide every child a risk-free, full-term pregnancy and a risk-free early childhood. A pregnancy must be free of alcohol, smoking, and drugs—the three greatest teratogens the world has ever known. We must guarantee adequate food, absence of abuse, and some level of enrichment for all children. There is no problem facing civilization that will not benefit from improving the lives and quality of our children. Are these all public health matters or can scientists have an influence through research, and not just as informed political activists? For example, can we develop an "Ames test" for teratogens, a simple assay that can be used to screen drugs for birth defects? We should be concerned with infant mortality, sudden infant death, and birth defects. What are the hormonal factors that regulate the length of pregnancy? If we knew, we might be able to prevent premature birth and the devastating problems that are exacerbated in low birth-weight infants.

Currently, I serve on the Council of the National Institute of Child Health and Human Development. This vantage point provides a panorama of our country's science directed to these "applied" problems. Most of the research on these subjects takes place in obstetrics-gynecology or pediatric departments of medical schools. My input at Council meetings is probably reminiscent of the phage molecular biologists trying to jump-start the embryologists 30 years ago. I can see the gap between the modern biology that we take for granted and the research carried out in these clinical departments. Yet medical schools are now establishing developmental biology departments, a practice unheard of even 10 years ago. They recognize the power of modern biology and look for guidance. Developmental biologists, who benefited so greatly from the disciplines and methods of the other life sciences, and from transfusions of physicists, biochemists, geneticists, and molecular biologists are no longer practicing outdated and insular science. The most exciting discoveries reported each week in leading journals are likely to be in developmental biology. Developmental biology, the beneficiary of cross-fertilization, must now consider what it can teach others. This may sound arrogant, but it is the historical way that science advances. Basic scientists invented genetic medicine, a field that barely existed when I was a medical student. They revolutionized immunology. They are having their impact in the neurosciences. There is not a field of medical research that will not gain by the application of methods we have come to view as ordinary. Let us begin to consider some of the great problems that to date have only concerned the pediatricians, obstetricians, and gynecologists. Let us see if we can give these fields the excitement and urgency that now surrounds medical genetics. This will be a major challenge for developmental biologists in the coming 50 years.

REFERENCES

Berleth T, Burri M, Thoma G, Bopp D, Richstein S, Frigerio G, Noll M, Nüsslein-Volhard C (1988): The role of localization of bicoid RNA in organizing the anterior pattern of the *Drosophila* embryos. EMBO J 7: 1749–1756.

Boveri Th (1902): Uber mehrpolige Mitosen als Mittel zur Analyse des Zellkerns. Verhandlungen der physikalisch-medizinischen Gesellschaft zu Würzbürg. Neue Folge 35: 67–90.

Brinster RL, Allen JM, Behringe RR, Gelinas RE, Plamiter RD (1988): Introns increase transcriptional efficiency in transgenic mice. Proc Natl Acad Sci USA 85: 836–840.

Brown DD (1984): The role of stable complexes that repress and activate eukaryotic genes. Cell 37: 359–365.

Gilbert W (1978): Why genes in pieces? Nature 271: 501.

Gudernatsch JF (1912): Feed experiments on tadpoles. I. The influence of specific organs given as food on growth and differentiation. A contribution to the knowledge of organs and internal secretion. Arch Entwicklungsmech. Organ. 35: 457–483.

Hadorn E (1966): Dynamics of determination. In: "Major Problems in Developmental Biology: 25th Symposium of the Society of Developmental Biology." New York: Academic Press, pp 85–104.

Nasmyth K, Seddon A, Ammerer G (1987): Cell cycle regulation of *SW15* is required for mother-cell-specific *HO* transcription in yeast. Cell 49: 549–558.

Oppenheimer JM (1966): The growth and development of Developmental Biology. In "Major Problems in Developmental Biology: 25th Symposium of the Society of Developmental Biology." New York: Academic Press, pp 1–25.

Ptashne M (1986): A Genetic Switch. Cell Press, pp 128.

Stent G (1968): That was the molecular biology that was. Science 160: 390–395.

Wang Z, Brown DD (1991): A gene expression screen. Proc Natl Acad Sci USA 88: 11505–11509.

Wilson, EB (1924): "The Cell in Development and Heredity." New York: The Macmillen Co., pp 1232.

Evolutionary Conservation of Developmental Mechanisms, pages 9–19

2. Expression and Developmental Regulation of the Maize *Spm* Transposable Element

J. A. Banks, P. Masson, and N. V. Fedoroff

Department of Botany and Plant Pathology, Purdue University, West Lafayette, Indiana 47907 (J.A.B.); Laboratory of Genetics, University of Wisconsin, Madison, Wisconsin 35706 (P.M.); Department of Embryology, Carnegie Institution of Washington, Baltimore, Maryland 21210 (N.V.F.)

INTRODUCTION

Suppressor-mutator (Spm) is a family of maize transposable elements consisting of both autonomous elements and non autonomous or defective elements. The autonomous *Spm* element contains all of the genetic information required both for transposition of the element from one position in the host chromosome to another, and for transposition of other elements within the same genome (reviewed in Fedoroff 1983, 1989b). A defective *Spm* element, on the other hand, transposes only in the presence of an autonomous element. All but one of the defective elements isolated thus far are internally deleted forms of the autonomous element (Masson et al., 1987; Tacke et al., 1986; T. Sullivan, personal communication). Irreversible inactivation of the *Spm* element, therefore, coincides with a structural change within the element itself. Other recent genetic and molecular studies have shown that the activity of the element can be heritably yet reversibly altered during the development of the maize plant. In this paper, we briefly describe the structure and function of the *Spm* element and review those studies that examine how the expression of the *Spm* element is controlled by the maize plant.

STRUCTURE OF THE AUTONOMOUS *Spm* ELEMENT AND ELEMENT-ENCODED TRANSCRIPTS

The autonomous, standard *Spm (Spm-s)* element is 8.3 kb long and contains two large open reading frames designated ORF1 and ORF2 (Fig. 1 Masson et al., 1987; Pereira et al., 1986). Several repetitive sequences occur at one or both ends of the element, including two 13-bp terminal inverted repeats (IRs), multiple copies of a conserved 12-bp sequence immediately adjacent to each IR, and multiple copies of a conserved 17-bp sequence within a region of the element encoding the first exon shared by all element-encoded transcripts (Masson et al., 1987). All of these repetitive sequences

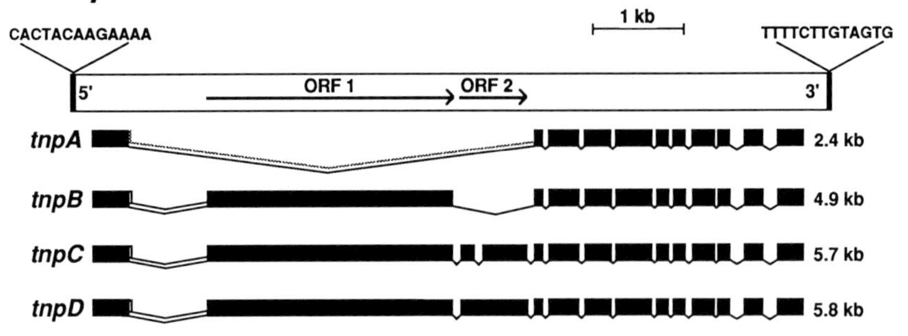

Fig. 1. The structure of the several large alternatively spliced *Spm* transcripts. The *Spm* element is represented by the box showing the two large open reading frames and terminal inverted repeats. The exons of the tnpA, tnpB, tnpC, and tnpD transcripts are shown on succeeding lines. The sizes of the transcripts, excluding the polyA tract, are shown at the right of the diagram.

are included in regions of the element which are thought to be important cis-determinants of *Spm* transposition frequency (Tacke et al., 1986; Masson et al., 1987; Schielfelbein et al., 1988).

The *Spm-s* element encodes multiple, alternatively spliced transcripts (Fig. 1). Four of these transcripts, designated tnpA, tnpB, tnpC, and tnpD, have been isolated and characterized from maize plants containing an *Spm-s* element. The tnpA and tnpD transcripts were also found in tobacco plants transformed with a functional *Spm-s* element (Masson et al., 1989). The most abundant transcript, tnpA, is 2.5 kb in length and contains 11 exons (Cuypers et al., 1988; Masson et al., 1989; Pereira et al., 1986). Both ORF1 and ORF2 sequences are removed (as intron 1) from the tnpA transcript by splicing. The tnpB, tnpC, and tnpD transcripts are 5–6 kb in length and are much less abundant than the tnpA transcript (Masson et al., 1989). These three transcripts all contain the tnpA exons plus 1, 2, or 3 additional exons derived from the intron 1 sequences of tnpA. TnpA and tnpB are monocistronic, while tnpC and tnpD are dicistronic (Fig. 2).

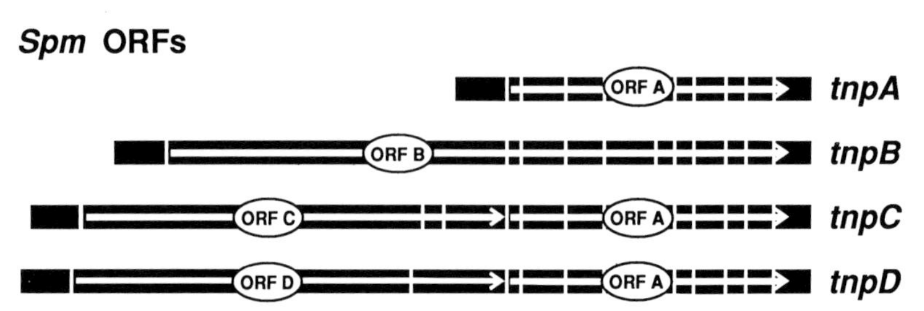

Fig. 2. A diagrammatic representation of the polypeptides encoded by the large ORFs of the *Spm* transcripts described in Figure 1. Each solid box represents an exon and each white arrow represents an ORF within the transcript.

To determine which of the four *Spm*-encoded transcripts are required for transposition of a *dSpm* element in trans, in vitro mutagenized *Spm* elements, as well as various combinations of tnpA, B, C, and D cDNAs, were tested for their ability to promote the transposition of a *dSpm* element in transgenic tobacco (Masson and Fedoroff, 1989; Masson et al., 1990; Periera and Saedler, 1989). The results of these experiments showed that tnpA and tnpD together are necessary and sufficient for the *Spm* transposase function. The tnpA gene product has been shown in vitro (Gierl et al., 1988b) and in vivo (Grant et al., 1990) to bind to a subset of the subterminal repetitive sequences, suggesting that tnpA is part of a complex involved in binding to the ends of the element during transposition. Although the function of the tnpD gene product is unknown, it is also likely to interact with the ends of the element given the significant homology between a domain of ORF1 (present in tnpD) and a domain within an ORF present in other plant transposable elements with terminal IRs nearly identical to those of *Spm* (Gierl et al., 1988b; Rhodes and Vodkin, 1988). It is not known at this time, however, if factors encoded by other transposable elements bind to the ends of their respective elements. In addition to its requirement for element transposition, tnpA may encode a positive autoregulatory gene product involved in the reactivation of an element in an inactive form as well as the coexpression of the element and the gene into which it is inserted in some *Spm* insertion alleles (Banks et al., 1988; Gierl et al., 1988a; Fedoroff and Banks, 1988; Masson et al., 1987, 1989). The tnpA gene product may also be required for the suppression of genes with some *dSpm* insertions (Grant et al., 1990).

THE PROGRAMMABLE AND STABLE FORMS OF THE *Spm* ELEMENT

In their studies of the *Spm* and closely related *Enhancer* (*EN*) elements, McClintock and Peterson isolated forms of the autonomous element whose phase of activity could be reversed in a characteristic and heritable pattern during the development of the plant, revealing that the expression of the element could be developmentally controlled by the plant (McClintock, 1958; Peterson, 1966; Fowler and Peterson, 1978). To understand the genetic and molecular basis of *Spm* inactivation and reactivation, and perhaps the developmental regulation of the element, derivatives of the *Spm* insertion allele *a-m2-7991A1* capable of undergoing changes in activity during the development of the maize plant were identified and characterized (Fedoroff and Banks, 1988; Masson et al., 1987). This allele contains an *Spm-s* insertion within the promoter of the maize *a* gene (Schwarz-Sommer et al. 1987); the *a* gene is required for anthocyanin synthesis in the plant. Kernels heterozygous for *a-m2-7991A1* (*Spm* active) and a recessive, mutant allele of *a* have a number of purple sectors on a palely pigmented background. Each purple sector represents a clone of cells derived from a single cell in which the element has been excised, restoring full *a* gene activity. The pale purple background results from a low but detectable level of *a* gene expression in cells where the active element remains inserted. Inactive element derivatives of the *a-m2-7991A1* allele are easily distinguished from the original *a-m2-7991A1* allele by phenotype. Regions of the kernel containing an inactive element derivative of the *a-m2-7991A1* allele are completely colorless. Inactive elements, unlike defective *Spm* elements which also give a colorless kernel phenotype, can be readily reactivated resulting in a mutable phenotype, at least in sectors of the kernel.

The genetic properties of the inactive element derivative were determined by monitoring the activity of the element in progeny kernels as the inactive element was transmitted from one generation to another (Fedoroff and Banks, 1988). Both the egg and pollen from main and tiller stalks were used as gamete donors of the element to the next generation. The heritability of the inactive state was initially shown to be low, but it increased after several generations of selection, to a point where almost all progeny kernels receiving an inactive element from its parent had an inactive element phenotype (Fedoroff and Banks, 1988). Thus, the first property of the inactive derivative is that its phenotype is heritable. The second property of the inactive element is that it can be reactivated during plant development in a very predictable way. The element has the highest probability of being reactivated when transmitted though a tiller gamete as compared with gametes produced on the main stalk of the same plant, and is least likely to be reactivated, or more likely to be inactivated, when transmitted though male gametes compared with female gametes of the same plant. A third property of the inactive element is that it can be reprogrammed during development. It was found, for example, that a colorless F1 kernel which receives an inactive *Spm* element through a FO tiller gamete in which the element has not been reactivated, tends to have a higher probability of being reactivated in the subsequent (F2) generation than if the F1 kernel receives the inactive *Spm* element through an FO main stalk gamete. In other words, even though an element may not be reactivated during development, it may be reprogrammed during development to favor reactivation in the next generation. The final property of the element is that it can be transiently reactivated in the presence of another *Spm* element (McClintock, 1958, 1959; Fedoroff and Banks, 1988). *Spm* elements for which these four properties hold are designated *programmable* because they display a variety of heritable differential patterns of programs of expression during plant development.

The expression of the original active *Spm* element in the *a-m2-7991A1* allele is considered stable because the probability that the activity of the element will change during development, and from generation to generation, is extremely low (Federoff and Banks, 1988). Similarly, stable inactive *Spm* elements, i.e., elements that have a very low probability of being expressed, have also been identified (Fedoroff, 1989a). The stable inactive, or cryptic,

Spm element can be genetically distinguished from a defective *Spm* element only by its ability to be reactivated by another active element in the same genome.

The reactivation of a cryptic element to an active form is progressive, requiring exposure to another active element for more than one generation (Fedoroff, 1989a; Banks et al., 1988). Before reaching a stably active state, the cryptic element passes through a number of intermediate steps where the activity of the element is unstable and may have one of several specific developmental programs of expression (Fedoroff, 1989a). The genetic behavior of all forms of the element indicate that the stably active and cryptic forms of the element are extremes in a continuum; between these two extremes are forms of the element in which the element's activity is developmentally determined. It was concluded from these genetic analyses that the mechanism that determines the activity of an *Spm* element has two genetically distinguishable components. One component, the phase setting, determines whether the element is genetically active or inactive. The other component, the phase program, determines the heritability of an element's setting and the timing and frequency of reversals of phase setting during the development of the plant (Fedoroff and Banks, 1988).

CYTOSINE METHYLATION AT THE 5′ END OF THE ELEMENT CORRELATES WITH ELEMENT INACTIVATION

Spm element methylation was analyzed in DNA isolated from plants having either a stably active element, a cryptic element, or an inactive programmable element (Banks et al., 1988). A correlation between the element's state of activity and methylation of cytosine residues within a 600-bp sequence at the 5' end of the element was observed. The methylatable cytosine residues tested outside of this region of the element were found to be extensively methylated regardless of the activity of the element (Banks et al., 1988). None of the flanking *a* gene sequences were found to be methylated (Banks et al., 1988).

Elements with an active phase setting could be distinguished from elements with an inactive phase setting (both cryptic and programmable) by the extent of methylation of cytosine residues between the left end of the element and the element's transcription initiation site at nucleotide position 209 (Banks et al., 1988; Pereira et al., 1986). Methylatable cytosine residues tested within this region of the element were not methylated if the element was active and were extensively methylated if the element was inactive. This region of the element, which contains the putative promoter of the element, has been designated the upstream control region, or UCR. Methylation of residues within the UCR also correlated with transcriptional inactivation of the element (Banks et al., 1988). Several *Spm*-homologous transcripts, present in plants containing active elements, were absent from plants containing cryptic or inactive programmable elements.

There are differences between the active, inactive programmable, and cryptic elements in the extent of element methylation of cytosine residues within a region immediately downstream of the UCR. This region of the element encodes the first untranslated exon of the *Spm* transcripts, is extremely rich in guanine and cytosine, containing many methylatable cytosine residues (Fig. 3) and, as previously mentioned, contains multiple copies of a conserved 17 bp sequence. Cryptic elements were found to be extensively methylated at all methylatable cytosine residues tested within this region. Active elements were found to be unmethylated, and inactive programmable elements were found to be partially methylated at the same sites. The extent of methylation within this region of the element, termed the downstream control region, or DCR, correlates with the heritability of the inactive state (Banks et al., 1988).

Element methylation was examined in plants containing inactive programmable elements while they were selected over a period of three generations for increased heritability of the inactive state (Banks and Fedoroff, 1989). An increase in the genetic stability of the inactive phase was shown to correlate with an increase in the fraction of elements methylated within the UCR and DCR. Complete methylation of the UCR and partial methylation of the DCR was observed in plants that transmitted the element in an inactive form to virtually all of its progeny if the main stalk tassel was used as the gamete source, yet was capable of differential reactivation during development. This suggests that methylation of the DCR, and not the UCR, is correlated with genetic stabilization of the inactive phase of the element (Banks and Fedoroff, 1989).

Changes in element methylation were also analyzed in plants containing either inactive programmable elements, or cryptic elements in the presence of other active elements (Banks et al., 1988). Genetic reactivation of a programmable inactive element by another active element was shown to correlate with the loss of methylation of tested sites within the reactivated element's UCR and DCR. Methylation of residues outside the UCR and DCR of the reactivated element did not change. Methylation of the UCR and DCR of a cryptic element in the presence of another active element for one generation was also observed to decrease. The exposure of a cryptic element to an active element for a single generation was not sufficient to reactivate the element; however, it did increase the probability of genetic reactivation of the element (Fedoroff, 1989a). Complete demethylation of the cryptic element's UCR and partial demethylation of the the cryptic element's DCR were observed only after the cryptic element had been exposed to an active element for two to three generations. These observations, together with the genetic observation that the conversion of an element in a cryptic form to an inactive programmable form is gradual, requiring more

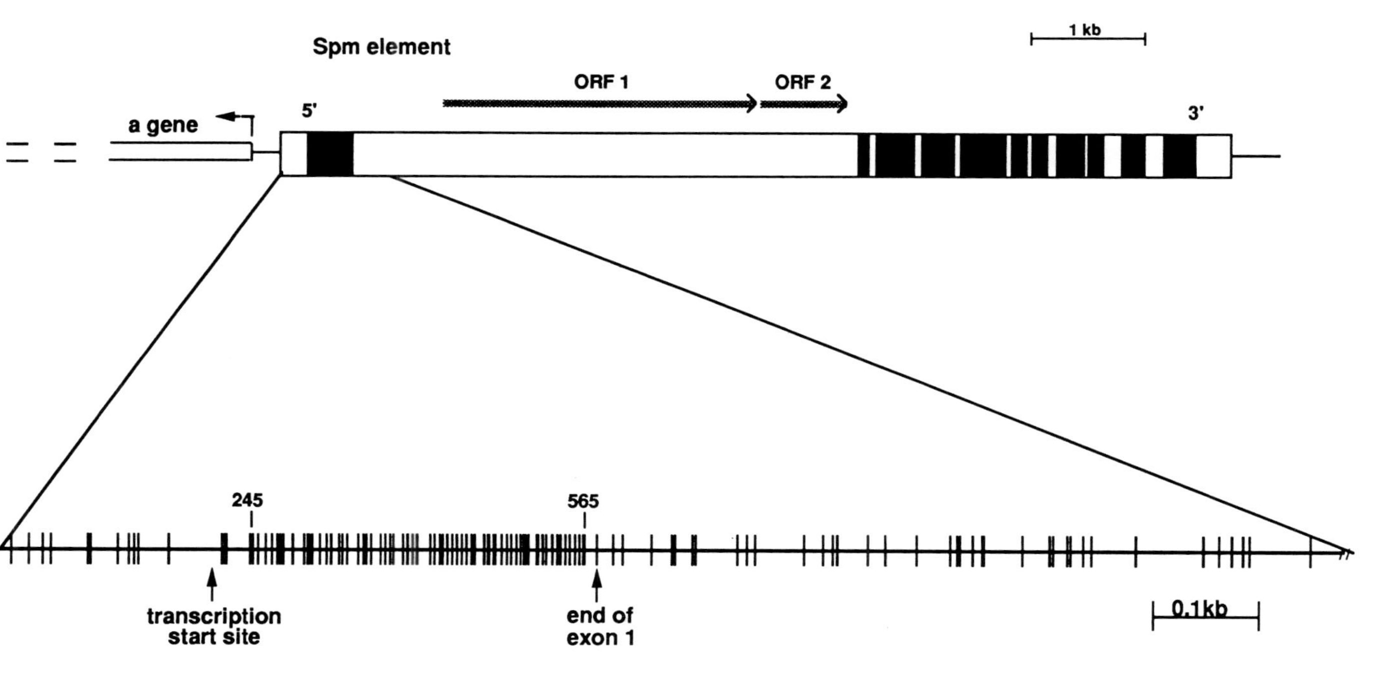

Fig. 3. A diagrammatic representation of *Spm* at a (top of figure) and the distribution of methylatable cytosine residues at the left end of the element (bottom of figure). The black boxes represent the exons of tnpA. The vertical tick marks represent cytosine residues within CG dinucleotides of CNG trinucleotides that may be methylated.

than one generation to achieve (Fedoroff, 1989b), also suggest that loss of DCR methylation is correlated with genetic destabilization of the cryptic element's program of expression.

Inactive programmable elements are also transcriptionally activated in the presence of another element (Banks et al., 1988). This suggests that *Spm* encodes a positive regulatory gene product that is capable of genetic and transcriptional activation of the inactive element by altering or interfering with methylation of the element's UCR and DCR (Fig. 4). As previously mentioned, this positive regulatory product is likely to be the tnpA gene product, although this has not yet been proven.

CHANGES IN ELEMENT METHYLATION DURING PLANT DEVELOPMENT

The genetic properties of the inactive programmable element suggest that the element is capable of reactivation during plant growth (Fedoroff and Banks, 1988). Developmental changes in the methylation of an inactive

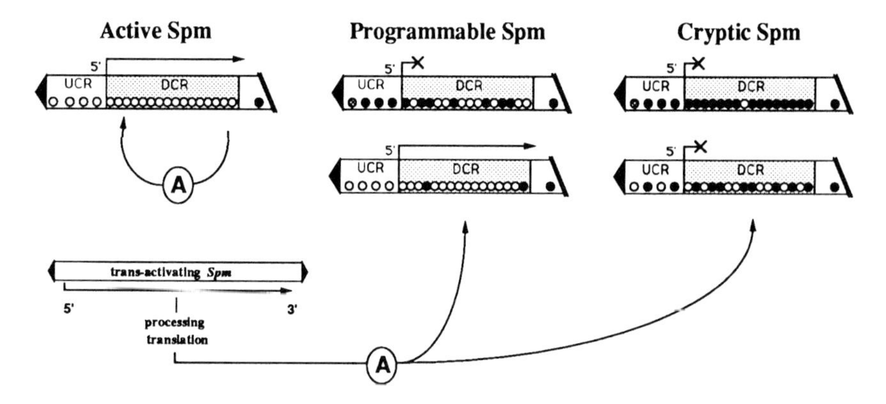

Fig. 4. A diagrammatic representation of *Spm* regulation. The boxes represent the *Spm* element's 5′ end. The element's transcription initiation site is marked by the base of the complete arrow (denoting transcriptional activity) or truncated arrow (denoting transcriptional inactivity). The upstream (UCR) and downstream (DCR) control regions are marked and circles within each, as well as downstream from the DCR, represent methylatable cytosine residues. Filled and open circles represent methylated and unmethylated sites, respectively. The hatched circle within the UCR denotes a site which is never fully methylated within a population of element in a plant. The box at the lower left represents a *trans*-activating *Spm* element and the circled A represents its positive regulatory gene product or products. In this diagram, the active *Spm* element is represented as producing its own regulatory gene product(s) which act to maintain the element in a transcriptionally active and hypomethylated state. The *programmable Spm* is represented as in a transcriptionally inactive form, except in the presence of a *trans*-activating *Spm*, which promotes its transcriptional activation and hypomethylation. The *cryptic Spm* is represented as transcriptionally silent regardless of the presence of a *trans*-activating element. However, the level of methylation of a *cryptic Spm* decreases in the presence of the *trans*-acting element. This, in turn, is correlated with an increase in the probability of genetic reactivation (see text).

programmable element were assessed by characterizing the methylation of such elements isolated from various plant tissues (Banks and Fedoroff, 1989). Dramatic differences in element methylation were evident in embryonic and endosperm tissues from mature, dormant kernels. The UCR and DCR were more extensively methylated in endosperm than in embryo tissues, regardless of whether the element was transmitted through the male or female germline to the kernels assayed. During vegetative growth of the main stalk, methylation of the UCR and DCR either remained constant or increased in progressively more apical leaves of the plant's main stalk. The UCR and DCR of elements isolated from tiller leaves of the same plants were not methylated. Clear differences in methylation were also observed in elements isolated from vegetative structures (leaves and ear husks) and reproductive structures (pollen and unfertilized ears). The UCR and DCR of elements were most extensively methylated when isolated from pollen, less extensively methylated if isolated from unfertilized ears, and least extensively methylated when isolated from leaves and ear husks of the same plant. The observed differences in methylation of elements in an inactive programmable form isolated from different parts of the plant suggest that there are at least three periods in maize development where changes in element methylation occur. The first occurs during a period between fertilization and development of the mature embryo where the inherited pattern of element methylation is at least partially erased. This erasure does not appear to occur in cell lineages giving rise to the mature endosperm of the same kernel. The second period occurs during vegetative growth of the primary (apical) meristem of the plant, where element methylation was observed either to remain constant or to increase in more apical leaves of the plant's main stalk. Similar increases in element methylation were not observed in elements isolated from the tiller stalks. The third period occurs during the development of the male and female reproductive tissues, when the element becomes extensively methylated. These results suggest that the molecular mechanism underlying element reprogramming and resetting during development involves the erasure of element methylation very early in the development of the embryo. The loss of element methylation that occurs during embryo development closely resembles the loss of element methylation that occurs after exposure of an inactive element to another active element. This coincidence suggests that other developmentally restricted factors not encoded by the element interact with the element's UCR and DCR to affect element methylation during embryogenesis. The phase setting and phase program of the element in germline tissues, which determines the phase program and phase setting in the zygote of the next generation, correlates with the extent of element remethylation that occurs during the growth of the plant. Furthermore, the extent of element methylation and the inactivity of the element in germline tissues correlates with the distance of the meristem producing the germline tissue from the base of the the plant (Fedoroff and Banks, 1988). Elements within the main stalk tassel, for example, are more likely to be methylated and inactive than elements of a tiller tassel which is derived from a meristem positioned below the main stalk tassel meristem.

SUMMARY

Recent studies have demonstrated that the *Spm* element is subject to inactivation by two mechanisms having very different consequences. The element may be irreversibly and permanently inactivated if certain sequences encoding the functional transposase are deleted from the element. The element can also be inactivated by a mechanism that is highly heritable yet reversible. This epigenetic mechanism can stably inactivate the element or can program the element to be differentially expressed during the development of the plant. Genetic and transcriptional inactivation of an element, as well as developmental programming of the element, is associated with differences in methylation of sequences surrounding the element's transcription initiation site. Genetic and transcriptional reactivation of an inactive element is associated with the presence of an *Spm*-encoded gene product(s) and other unidentified plant-encoded factors that can modulate *Spm* element methylation during development.

REFERENCES

Banks J, Masson P, Fedoroff N (1988): Molecular mechanisms in the developmental regulation of the maize *Suppressor-mutator* transposable element. Genes Dev 2:1364–1380.

Banks J, Fedoroff N (1989): Patterns of developmental and heritable changes in methylation of the *Suppressor-mutator* transposable element. Dev Genet 10:425–437.

Cuypers H, Dash S, Peterson P, Saedler H, Gierl A (1988): The defective En-1102 element encodes a product reducing the mutability of the *En/Spm* transposable element system of *Zea mays*. EMBO J 7:2953–2960.

Fedoroff N (1983): Controlling elements in maize. Shapiro J (ed): In "Mobile Genetic Elements." New York: Academic Press, pp 1–63.

Fedoroff N (1989a): The heritable activation of cryptic *Suppressor-mutator* elements by an active element. Genetics 121:591–608.

Fedoroff N (1989b): Maize transposable elements. In Hoew, M Berg D (eds): "Mobile DNA." Washington: American Society for Microbiology, pp 375–411.

Fedoroff N, Banks J (1988): Is the *Suppressor-mutator* element controlled by a basic developmental regulatory mechanism? Genetics 120:559–577.

Fowler R, Peterson P (1978). An altered state of a specific *En* regulatory element induced in a maize tiller. Genetics 90:761–782.

Gierl A, Cuypers H, Lutticks S, Pereira A, Schwarz-Sommer Z, Dash S, Peterson P, Saedler H (1988a): Structure and function of the *En/Spm* transposable element system in *Zea mays*: Identification of the suppressor component of *En*. In Nelson O (ed): "Plant Transposable Elements." New York: Plenum Press, pp 115–119.

Gierl A, Lutticke S, Saedler H (1988b): tnpA product encoded by the transposable element *En-1* of *Zea mays* is a DNA-binding protein. EMBO J 7:4045–4053.

Grant S, Gierl A, Saedler H (1990): *En/Spm*-encoded tnpA protein requires a specific target sequence for suppression. EMBO J 9:2029–2035.

Masson P, Fedoroff N (1989): Mobility of the maize *Suppressor-mutator* element in transgenic tobacco cells. Proc Natl Acad Sci USA 86:2219–2223.
Masson P, Rutherford G, Banks J, Fedoroff N (1989): Essential large transcripts of the maize *Spm* transposable element are generated by alternative splicing. Cell 58:755–765.
Masson P Strem M, Fedoroff N (1991): The tnpA and tnpD gene products of the *Spm* element are required for transposition in tobacco. Plant Cell 3:73–85.
Masson P, Surosky R, Kingsbury J, Fedoroff N (1987): Genetic and molecular analysis of the *Spm*-dependent *a-m*2 alleles of the maize *a* locus. Genetics 177:117–137.
McClintock B (1958): The *Suppressor-mutator* system of control of gene action. Carnegie Inst Wash Year Book 57:415–429.
McClintock B (1959): Genetic and cytological studies of maize. Carnegie Inst Wash Year Book 58:452–456.
Pereira A, Cuypers H, Gierl A, Schwarz-Sommer A, Saedler H (1986): Molecular analysis of the *En/Spm* transposable element system of *Zea mays*. EMBO J 5:835–841.
Pereira A, Saedler H (1989): Transpositional behavior of the maize *En/Spm* element in transgenic tobacco. EMBO J8:1315–1321.
Peterson P (1966): Phase variation of regulatory elements in maize. Genetics 54:249–266.
Rhoades M, Vodkin L (1988): A conserved Zn finger domain in higher plants. Plant Mol Biol 12:593–594.
Schiefelbein J, Raboy V, Kim H, Nelson O (1988): Molecular characterization of *Suppressor-mutator (Spm)*-induced mutations at the *bronze-1* locus in maize: The *bz-m13* alleles. In Nelson O (ed): "Plant Transposable Elements." New York: Plenum Press, pp 261–278.
Schwarz-Sommer Z, Shepherd N, Tacke E, Gierl A, Rohde W, LeClercq L, Mattes M, Berndtgen R, Peterson P, Saedler H (1987): Influence of transposable elements on the structure and function of the *A*1 gene of *Zea mays*. EMBO J 6:287–294.
Tacke E, Schwarz-Sommer Z, Peterson P, Saedler H, (1986): Molecular analysis of states of the *A* locus of *Zea mays*. Maydica 31:83–91.

Evolutionary Conservation of Developmental Mechanisms, pages 21–38

3. The Chicken Antibody Repertoire Is Generated by Gene Conversion During B Cell Development in the Bursa of Fabricius

Elizabeth A. Hurley, Larry W. Tjoelker, and
Craig B. Thompson

Howard Hughes Medical Institute and Departments of Internal Medicine and Microbiology/Immunology, University of Michigan Medical Center, Ann Arbor, Michigan 48109

INTRODUCTION

The ability of the vertebrate immune system to specifically recognize a wide variety of antigens has provided an intriguing developmental puzzle. Development of an effective circulating immune repertoire requires that vertebrates produce an estimated 10^9 distinct antibody molecules. Yet the vertebrate genome is not large enough to encode individually this large a number of different antibody molecules. Functional immunoglobulin genes are instead created by a series of recombination events that occur during B cell development (reviewed in Tonegawa, 1983). Specifically, the antigen binding exon of both the immunoglobulin heavy- and light-chain genes are formed through the recombination of individual elements from families of functional variable (V), diversity (D), and joining (J) gene segments. In humans and mice, Ig diversity is generated by differences in the combinatorial associations of these elements, along with variations in the precise position of the joining and de novo nucleotide synthesis at the joint between the individual elements (Alt and Baltimore, 1982).

Not all vertebrates, however, use a strategy of recombinatorial diversity to create a primary immunologic repertoire. Chickens have been shown to possess single functional V and J gene segments capable of rearrangement within both their heavy (IgH) and light (IgL) chain immunoglobulin (Ig) loci (Reynaud et al., 1985, 1989). In birds, a diverse Ig repertoire is generated as a consequence of B cell development within a novel organ, the bursa of Fabricius. In this manuscript, we will summarize the current understanding of the molecular events associated with avian B cell development, focusing on the mechanism by which diversity is created within the immunoglobulin gene loci, and the role of the bursa of Fabricius in this process.

BURSA OF FABRICIUS, AN ESSENTIAL ORGAN FOR B CELL DEVELOPMENT IN AVIAN SPECIES

Avian species differ from mammals in possessing two organs, the thymus and the bursa of Fabricius, which are required for lymphoid development (Fig. 1). Each organ is necessary for the development of one of the two major lymphoid lineages. In the chicken, the thymus is localized as seven lobes on either side of the trachea, and its presence is essential for normal T cell development and function. Aspects of avian T cell development and its comparison to mammalian T cell development have recently been reviewed by Cooper et al. (1991), and will not be covered further in this review.

The other organ which is essential for lymphoid development in avian species is the bursa of Fabricius, which is located in the posterior cloaca, the common urogenital organ of the chicken. The bursa of Fabricius first begins to differentiate at 60 hours of embryonic development as a posterior invagination of the cloacal wall (Moore and Owen, 1966). Within the lumen of this invagination, specialized collections of stromal cells begin to develop between the epithelium and the basement membrane. These cells organize into approximately 10^4 individual follicular structures that begin to be

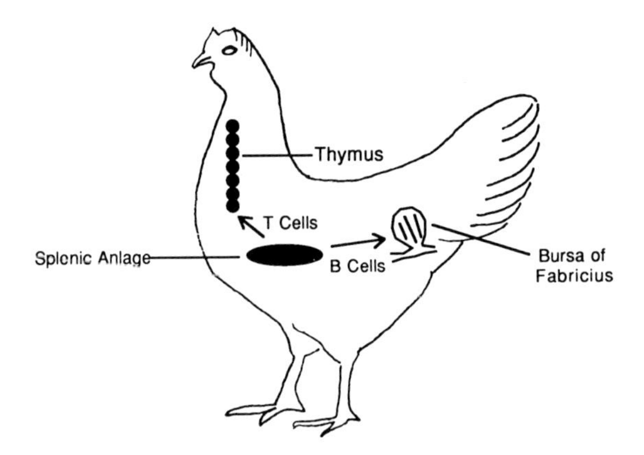

Fig. 1. The avian immune system. Lymphoid progenitors present in the splenic anlage migrate to the thymus and the bursa of Fabricius. T cell development occurs within the thymus, which is localized as seven lobes on either side of the trachea. B cell progenitors within the splenic anlage undergo immunoglobulin gene rearrangement and migrate to the bursa of Fabricius, the essential organ for B lymphocyte development. Within the bursa, Ig-positive cells proliferate and undergo somatic diversification of their Ig loci. The expanded pool of B cells, with diverse antigen specificities, then migrates to the periphery and constitutes the humoral immune response of the bird.

seeded by lymphoid cells, which migrate from the avian splenic anlage, beginning around day 10 of embryogenesis (Houssaint et al., 1976; Olah and Glick, 1978) (Fig. 2). The number of lymphoid cells within an individual follicle increases progressively until the birds are 2–4 weeks of age, at which time an individual follicle has approximately 10^5 lymphocytes present within the medullary space, and lymphocytes can be observed leaving the organ through the capillary network in the cortical region of the follicle (Grossi et al., 1976; Lydyard et al., 1976). At 4–6 weeks of age, the bursa of Fabricius can contain up to 2×10^9 lymphocytes, making it one of the largest organs present in the bird at this stage of development (Olah and Glick, 1978). The bursa of Fabricius remains a dominant lymphoid organ until the bird reaches sexual maturity at 4–6 months of age; then the bursa undergoes involution (Glick, 1977). Only a fibrotic remnant can be observed in adult birds.

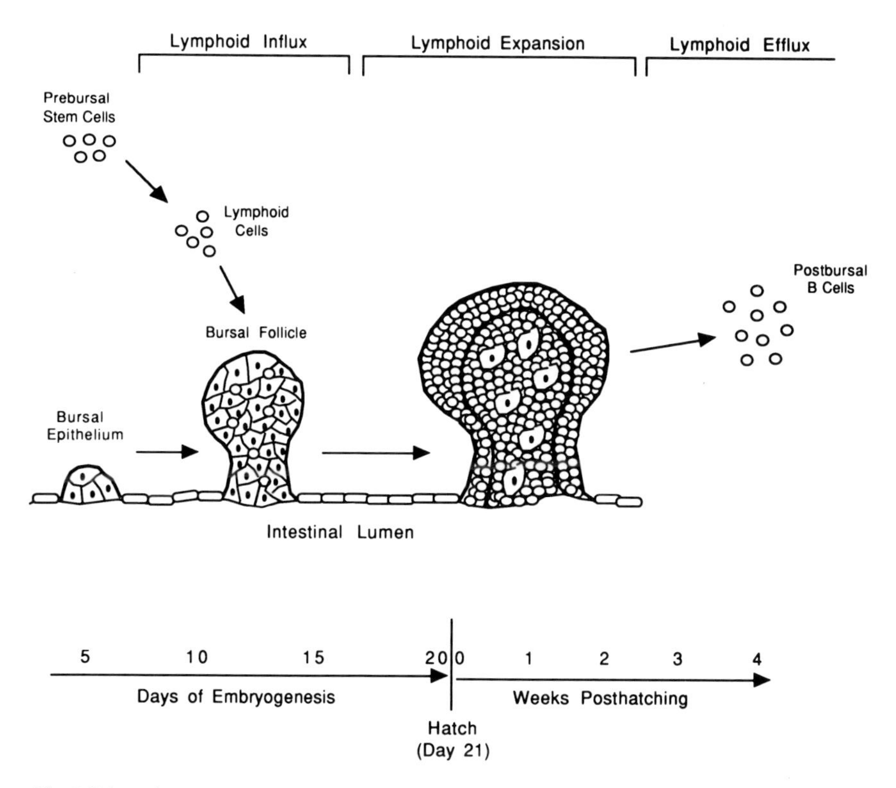

Fig. 2. B lymphocyte development within the bursa of Fabricius. Lymphoid cells migrating from the splenic anlage begin to seed bursal follicles at 10 days of embryogenesis. B lymphocyte proliferation occurs within the bursa, and the expanded B cell population migrates to the periphery via the capillary network in the cortex of the follicles.

Bursectomy experiments first performed by Glick and his colleagues (Glick et al., 1956) and subsequently by many other investigators (Cooper et al., 1969; Huang and Dreyer, 1978; Jalkanen et al., 1983 a,b) have demonstrated the importance of the period of B cell development in the bursa of Fabricius for generating a humoral immune repertoire. Animals bursectomized between days 17 and 18 of embryogenesis fail to develop either circulating antibodies or peripheral B cells (Cooper et al., 1969; Lerman and Weidanz, 1970). This finding suggests that at this point of development all the cells destined to become part of the B cell lineage are present within the bursa of Fabricius. Bursectomy after 4 weeks of age has no significant effect on the avian humoral immune system. However, bursectomy at progressively earlier timepoints during B cell development leads to limitation of the humoral immune repertoire as measured by the ability of the bird to respond to a variety of complex antigens (Granfors et al., 1982; Jalkanen et al., 1983 a,b; Eerola, 1984). These early experiments led to the suggestion that the bursa of Fabricius is the site of creation of the immunoglobulin repertoire in avian species.

RECOMBINATION IS CAPABLE OF GENERATING A SINGLE IMMUNOGLOBULIN MOLECULE

Studies of the generation of Ig diversity in mammals suggested a paradigm in which the primary Ig repertoire of all vertebrates would be created in large part by diverse recombination of many V, D, and J segments. It was presumed that in avian species rearrangement of these genes would occur in the bursa of Fabricius. This model came into question, however, upon the elucidation of the molecular organization of the Ig genes of the chicken by Reynaud et al. (1985, 1989) (Fig. 3). Their studies demonstrated that, while there is some diversity in the D regions of the heavy chain, both the Ig heavy and light chain genes contain single V (V_H and V_L) and J (J_H and J_L) elements capable of rearranging. Multiple V gene segments are present upstream of the single functional V genes in both the heavy and light chain loci. While these V gene segments share a high degree of sequence homology with the functional V_H and V_L, all are pseudogenes with molecular defects rendering them incapable of recombination (Reynaud et al., 1985, 1987, 1989). Thus, in the chicken, Ig rearrangement is not capable of generating a diverse Ig repertoire. It has also become apparent that Ig recombination occurs prior to B cell migration to the bursa of Fabricius (McCormack et al., 1989a; Mansikka et al., 1990). These findings indicated that the previously accepted hypothesis concerning the generation of an Ig repertoire in avian species must be altered. Much data has recently been generated in order to elucidate the molecular and cellular events which occur during the development of the avian humoral immune response.

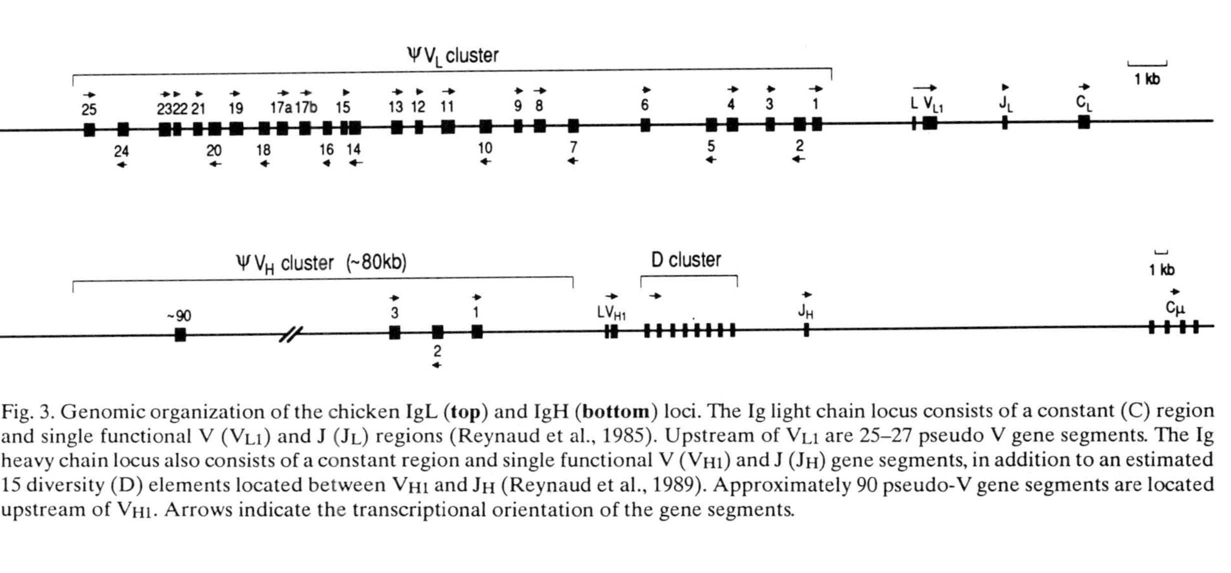

Fig. 3. Genomic organization of the chicken IgL (**top**) and IgH (**bottom**) loci. The Ig light chain locus consists of a constant (C) region and single functional V (V_{L1}) and J (J_L) regions (Reynaud et al., 1985). Upstream of V_{L1} are 25–27 pseudo V gene segments. The Ig heavy chain locus also consists of a constant region and single functional V (V_{H1}) and J (J_H) gene segments, in addition to an estimated 15 diversity (D) elements located between V_{H1} and J_H (Reynaud et al., 1989). Approximately 90 pseudo-V gene segments are located upstream of V_{H1}. Arrows indicate the transcriptional orientation of the gene segments.

IN AVIAN SPECIES IMMUNOGLOBULIN RECOMBINATION SERVES ONLY TO ACTIVATE IMMUNOGLOBULIN GENE EXPRESSION

The molecular mechanism by which avian B cells undergo immunoglobulin recombination is much the same as that which occurs in mammalian species. For example, in the Ig light chain (IgL) gene, the germline-encoded V (V_L) and J (J_L) segments that undergo recombination are flanked by recombination signal sequences made up of a heptamer segment, a 12- or 23-base spacer sequence, and a conserved nonamer segment (Reynaud et al., 1985). During recombination, a coding joint is formed between the V and J segments that leads to the production of an antigen-binding exon. A byproduct of this recombination process is the formation of a signal joint in which the two heptamer elements of the recombination signal sequences are precisely joined. In the avian B cell, this signal joint, containing the DNA between V and J, is deleted from the genome and can be detected as a circular episome which is present only transiently in cells following recombination (McCormack et al., 1989b) (Fig. 4). By following the presence of the recombination signal sequence episome during B cell development, it was shown that the recombination of both heavy and light chain genes occurs in hematopoietic cells that reside in the splenic anlage between days 10 and 15 of embryogenesis (Pickel et al., manuscript in preparation). These cells with rearranged Ig genes then migrate to the bursa of Fabricius to colonize the lymphoid compartment. Thus, in avian cells, Ig recombination serves primarily to activate transcription of the two immunoglobulin genes.

WITHIN THE BURSA OF FABRICIUS, THE SURFACE IMMUNOGLOBULIN-POSITIVE B CELLS UNDERGO PROLIFERATIVE EXPANSION

Immunoglobulin recombination in avian species is completed by day 15 of embryogenesis, at which point fewer than 1% of the total cells in the bursa of Fabricius are lymphocytes. However, over the ensuing week, there is an exponential increase in the number of lymphoid cells, so that by the day of hatch there are close to 10^8 lymphocytes present in the bursa of Fabricius. This increase in B cell number within the bursa of Fabricius arises primarily from proliferation within the organ (Lydyard et al., 1976). Virtually all bursal B cells are in exponential growth during this period of development, and the increase in B cell number cannot be generated by the migration of progenitor cells that have newly recombined their immunoglobulin loci. Entry into the proliferative phase of B cell development within the bursa of Fabricius requires surface expression of an Ig molecule composed of the germline-encoded V and J gene segments (McCormack et al., 1989a). Random recombination of two segments such as the V and J gene segments leads to an out-of-frame joint two-thirds of the time. While B cells with such Ig rearrangements appear to be able to migrate to the bursa of Fabricius, they are not capable of entering the proliferative phase of B cell development within the bursa of Fabricius (McCormack et al., 1989a). Therefore, B cells are selected in the bursa for the presence of a productive Ig rearrangement. By 18 days of development, over 94% of the recombined immunoglobulin gene segments that are present in the bursa of Fabricius are in-frame. The 5–6% of alleles which are out-of-frame appear to be present in B cells that have rearranged both parental Ig alleles (McCormack et al., 1989a). These findings suggest that it is the ability of the B cell to express a surface immunoglobulin molecule that induces entry into the proliferative phase of development in the bursa of Fabricius. The expanding B cell population within the bursa subsequently migrates to the periphery, as shown in Figure 4. Expansion of Ig-positive B cells within the bursa generates the large number of B cells needed to constitute the avian humoral immune system, and thus compensates for the limited number of cells that undergo immunoglobulin recombination early in development.

AN IMMUNOGLOBULIN REPERTOIRE IS CREATED DURING B CELL DEVELOPMENT IN THE BURSA OF FABRICIUS

The above studies demonstrated the importance of Ig recombination in the generation of a large number of B cells that could contribute to the humoral immune system of the chicken. However, subsequent to the rearrangement event, somatic diversification of recombined immunoglobulin genes must take place during B cell development in order to generate antibodies recognizing a multitude of diverse antigens. Therefore, rearranged immunoglobulin V gene segments present in IgL were examined for evidence of sequence variation during B cell development within the bursa of Fabricius (Thompson and Neiman, 1987). As seen in Figure 5, in the chicken SC strain there is a Sca I site present within the rearranged V_{L1} gene segment. However, Southern analysis of the rearranged V_{L1} gene segments within developing bursal B cells demonstrated progressive loss of this Sca I site as the B cells proliferate. Additional restriction endonuclease sites present within the rearranged V gene segments were likewise altered. These findings indicate that the primary V–J rearrangement is undergoing modification as the B cell population expands within the bursa. Surprisingly, however, restriction site diversification was not observed in the leader exon or within the J gene segment. The diversification process is also restricted to the recombined V gene segments, as sequence diversification was not observed on the V_{L1} gene segment present on the unrearranged parental allele in the same bursal lymphocytes. Therefore, a novel form of somatic diversification occurs within the recombined immunoglobulin V gene segments during B cell development in the bursa of Fabricius.

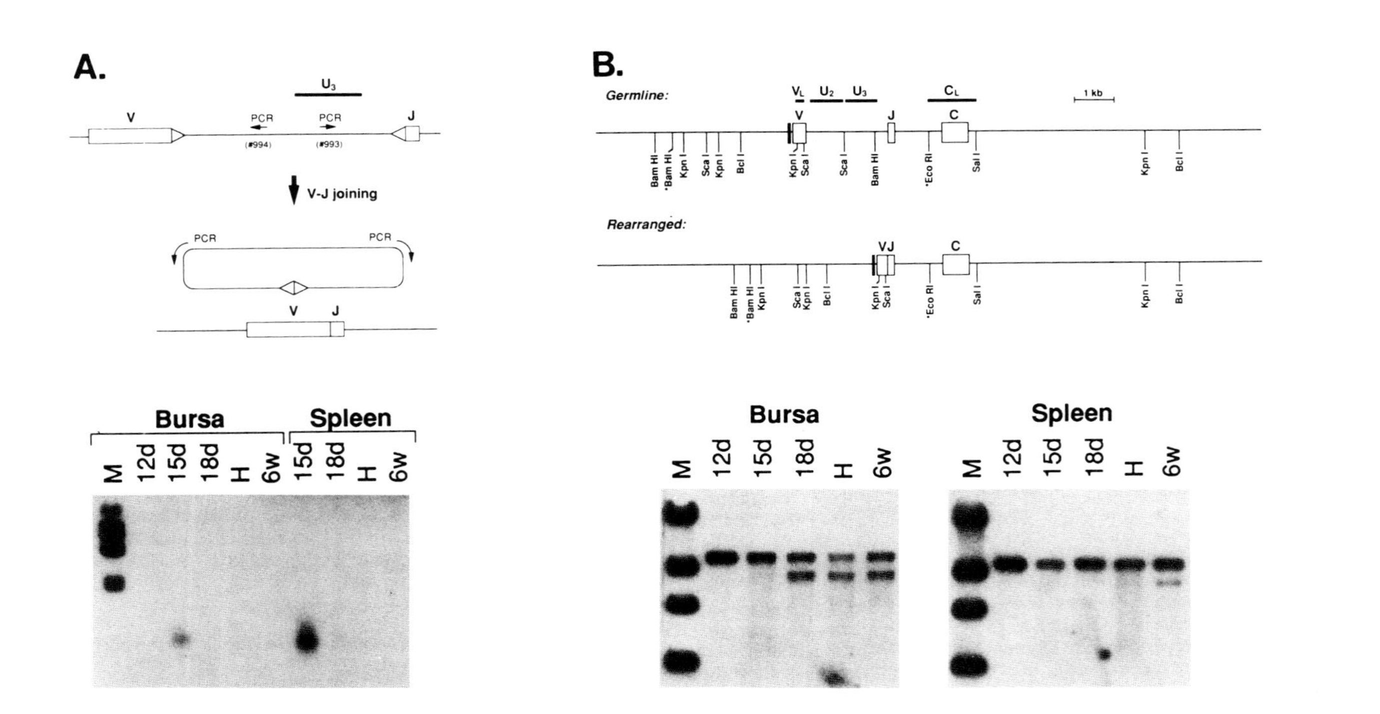

A.
U3
V
PCR
PCR
J
(#994)
(#993)
V-J joining
PCR
PCR
V
J
Bursa
Spleen
M
12d
15d
18d
H
6w
15d
18d
H
6w
B.
Germline:
VL
U2
U3
CL
1 kb
V
J
C
Bam HI
*Bam HI
Kpn I
Sca I
Kpn I
Bcl I
Kpn I
Sca I
Sca I
Bam HI
*Eco RI
Sal I
Kpn I
Bcl I
Rearranged:
VJ
C
Bam HI
*Bam HI
Kpn I
Sca I
Kpn I
Bcl I
Kpn I
Sca I
*Eco RI
Sal I
Kpn I
Bcl I
Bursa
M
12d
15d
18d
H
6w
Spleen
M
12d
15d
18d
H
6w

In order to determine the nature of this diversification, cDNAs were isolated from bursal lymphocytes at different stages of development by Reynaud and her colleagues (Reynaud et al., 1987). They were able to show that V gene segments of immunoglobulin light chain cDNAs undergo an extensive sequence diversification in the lymphocytes developing during the proliferative phase of B cell development in the bursa of Fabricius. By sequence analysis, it was demonstrated that the sequences acquired within the rearranged V_{L1} gene segments have precise templates within the pseudo-V(φV) gene segments present upstream of V_{L1}. This finding suggested that sequence substitution events occur between the functional V gene segment, and sequences present within the upstream pseudogene cluster present in both the heavy and light chain loci. As seen in Figure 6, rearranged V_L–J_L segments present in bursal lymphocytes on the day of hatching have undergone extensive tracts of such sequence substitution derived from φV_L gene segments. An individual rearranged V gene segment can undergo multiple rounds of these substitution events so that in the mature bird, rearranged V genes may contain sequences from up to ten different φV gene segments (Fig. 7).

SOMATIC DIVERSIFICATION RESULTS FROM INTRACHROMOSOMAL GENE CONVERSION

The mechanism by which sequence transfer between rearranged V genes and flanking pseudogenes occurs during B cell development in the bursa of Fabricius has been studied in considerable detail over the last several years, and is summarized in Figure 8. The diversification process was analyzed by Carlson et al. (1990) in an F1 cross of two inbred chicken strains which contain numerous polymorphisms throughout the immunoglobulin light chain gene

---

Fig. 4. Bursal B cells with rearranged Ig loci migrate to the periphery. **A**: **(Top)** V–J rearrangement in the chicken IgL locus results in production of a V–J coding joint, with deletion of a signal joint episome consisting of the DNA between the V and J coding segments. The signal joint episome can therefore be used as a marker for the presence of ongoing rearrangement in a population of cells. PCR primers were designed to amplify this signal joint episome within the bursa and spleen. PCR products were then electrophoresed in agarose gels, blotted onto nitrocellulose, and hybridized to the U3 probe. **(Bottom)** At 15 days of embryogenesis the signal joint episome can be detected only at low levels in the bursa, but at high levels in the spleen. Subsequent experiments have shown that the low level of episomes present in the bursa at day 15 is due to migration of B cells containing the episome from bursa to spleen. **B**: **(Top)** Organization of the germline and rearranged IgL locus. Bcl I digestion of the germline IgL locus yields a restriction fragment of 10 kb which hybridizes to the C_L probe upon Southern blotting. Bcl I digestion of the rearranged IgL locus yields a restriction fragment of 8 kb which hybridizes to the C_L probe. **(Bottom)** Southern blots of Bcl-I digested DNA from bursa and spleen hybridized to the C_L probe. Between days 15 and 18 of embryogenesis, a large increase in the number of rearranged Ig alleles is detected in the bursa. This increase results from proliferation of Ig-positive B cells in the bursa. B cells with rearranged Ig alleles migrate to the spleen where they can be detected at 6 weeks posthatching. H=hatching.

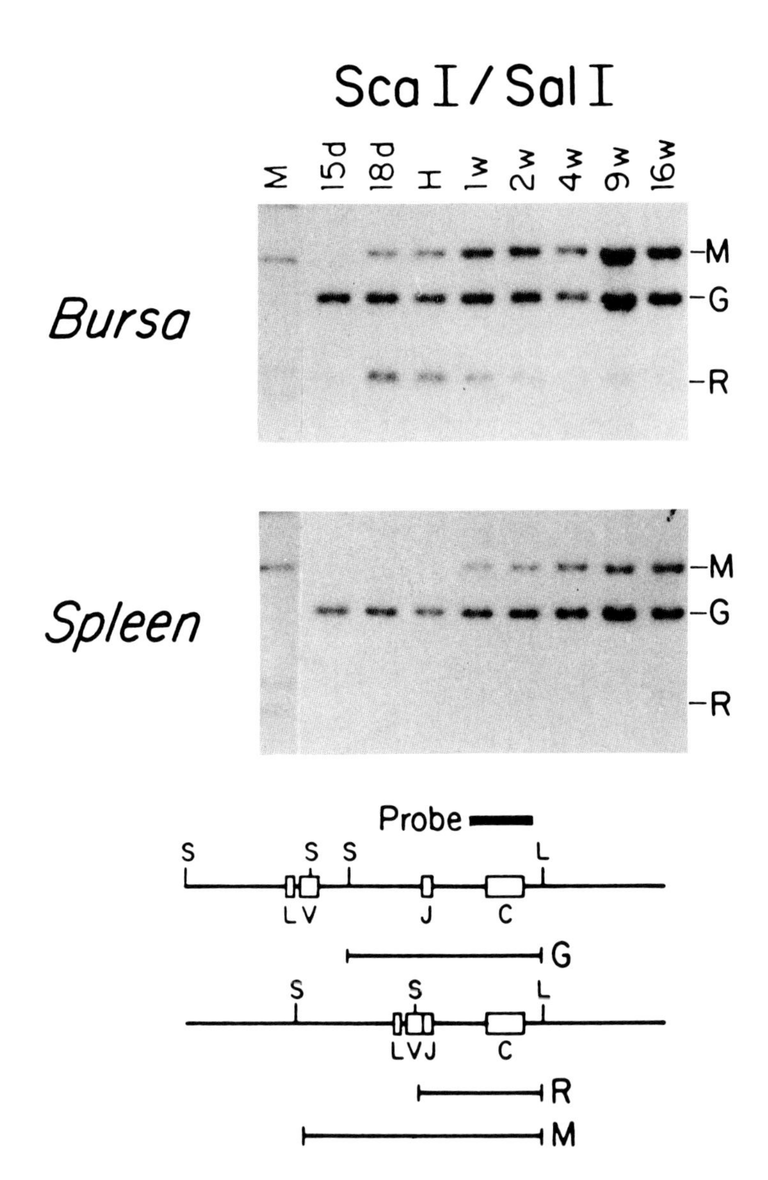

Fig. 5. Somatic diversification of rearranged IgL alleles in B lymphocytes occurs during B cell development in the bursa of Fabricius. Lymphocyte DNA from bursa and spleen at various stages of development was digested with the restriction enzymes Sca I (S) and Sal I (L) and hybridized to a probe specific for the C_L region of chicken IgL. As shown schematically at the **bottom** of the figure, germline (G), rearranged (R), and somatically modified (M) IgL alleles produce C_L-hybridizing restriction fragments of distinct sizes upon Sca I and Sal I digestion. As shown in the Southern blots (**Top**), in the bursa, rearranged alleles (R) are progressively somatically modified (M) as the B cells proliferate, from day 18 of embryogenesis onward, until at 16 weeks posthatching no unmodified rearranged alleles can be detected. Only B cells with Ig genes which have undergone somatic diversification migrate to the spleen, where they can be detected by 1 week posthatching. H=hatching.

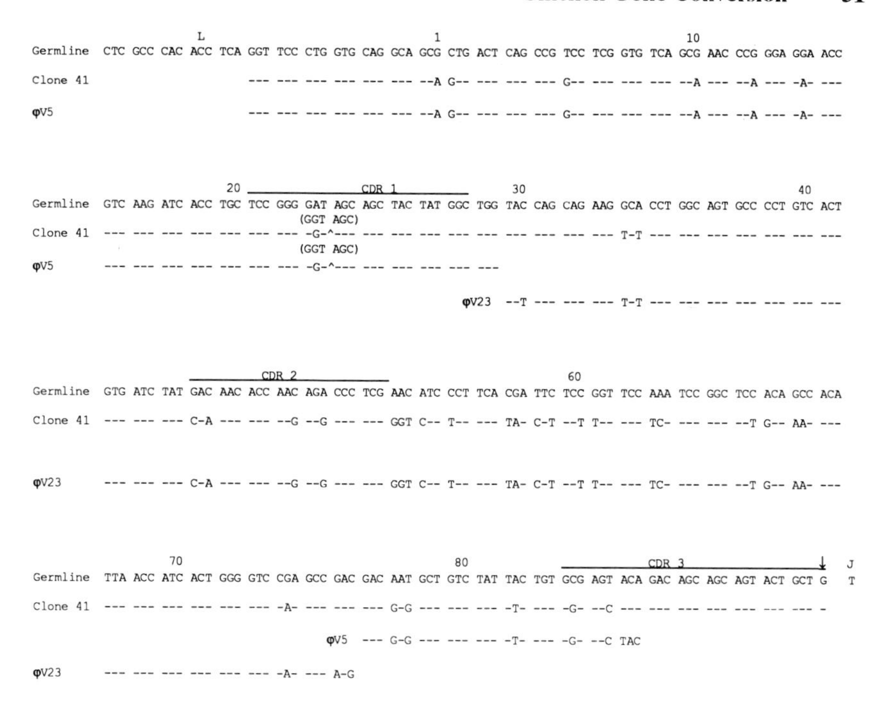

Fig. 6. Sequence comparison of the germline V_{L1} allele (top line) and the diversified V allele isolated from clone 41, a bursal lymphocyte cell line (second line). The diversified clone 41 allele is created by sequence substitution from φV5 (third line) and φV23 (bottom line); positions of the sequence substitutions are shown. All sequence substitution is within the rearranged V segment; no diversification is present within the leader (L) or J regions. Dashes indicate identity to the germline sequence.

locus. In this study, it was demonstrated that sequence transfer always occurs in an unidirectional fashion from the φV gene segments to the functional V gene segment. Sequence transfer was never observed to occur from the functional V gene segments to the pseudogenes, nor were the pseudogene elements capable of contributing sequence information to each other. Furthermore, all sequence transfer that could be characterized was always transferred from φV gene segments present in *cis* with a rearranged V gene segment. Using a variety of molecular strategies, no evidence was found for the transfer of sequence information in *trans* between parental alleles. In cell lines derived from bursal B cells, sequence diversification of the pseudogene segments was never detected in either a reciprocal or nonreciprocal fashion during B cell development within the bursa of Fabricius. Therefore, the primary mechanism of sequence diversification during B cell development in the bursa of Fabricius is intrachromosomal gene conversion.

McCormack and Thompson (1990) extended these observations by characterizing individual gene conversion events in greater molecular detail. This study confirmed that gene conversion events occur exclusively in *cis*,

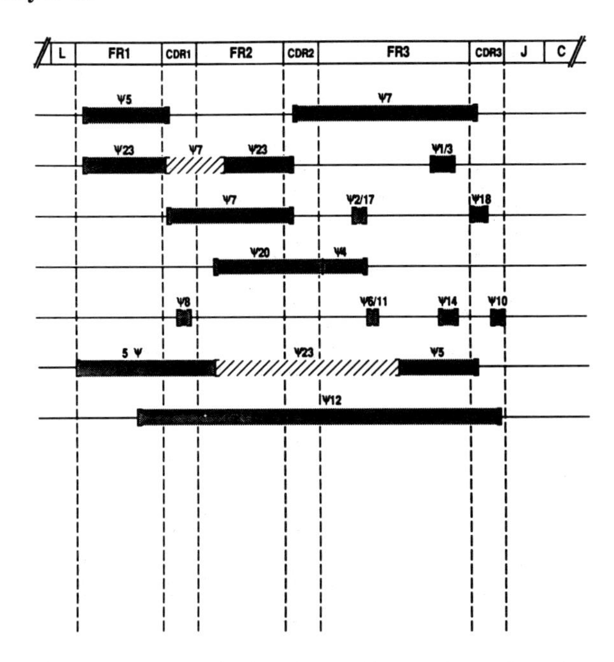

Fig. 7. Sequence substitutions detected in somatically diversified VL gene segments at 18 days of embryogenesis. The top line represents the germline IgL gene, with the V_L region composed of relatively conserved framework (FR) and more variable complementarity determining (CDR) regions. Blocks of sequence substitutions within V_{L1}, and the pseudo-V gene segments from which they are derived, are shown for seven isolated V_L alleles. The length, location, and number of sequence substitutions varies for each of these diversified alleles. At later time points in development, rearranged V genes can contain sequences from up to ten different φV gene donors.

and determined that the frequency with which a pseudogene is used as a donor is determined by a number of different criteria. Specifically, φV segments that are in close proximity to V_{L1} and that are transcriptionally oriented in the antisense direction are frequent donors. The extent of homology between the functional V and φVs is also a determinant of sequence transfer, as those φV genes with a greater length of homology are more likely to be sequence donors. These authors also found evidence that gene conversion occurs in a directional manner. In their data set, the 5′ ends of gene conversion events always began in a region of sequence identity between the rearranged V gene segment and the pseudogene donor templates. There also appears to be a requirement for a stretch of homology between the donor and the recipient at the 5′ end of events, as a first 45-base pair homology between the donor and recipient gene segments was never involved in a gene conversion tract. In contrast, 10–15% of 3′ ends

Gene Conversion

A.

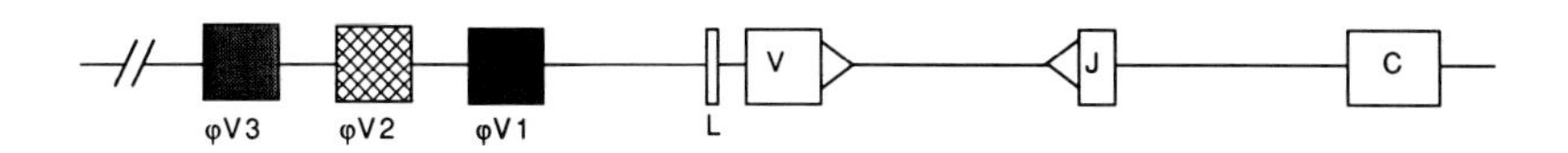

B.

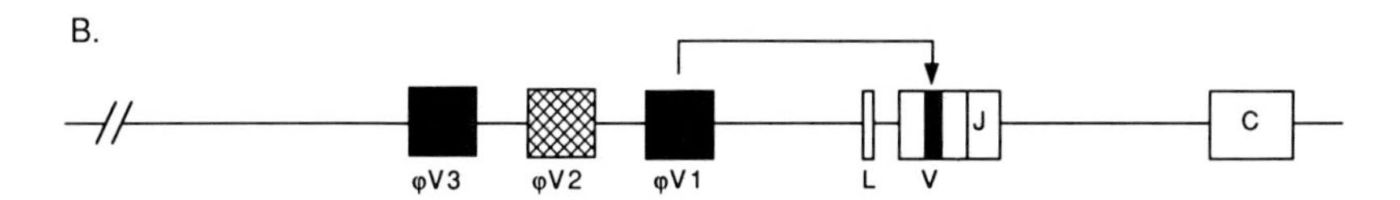

Fig. 8. Gene conversion of the chicken IgL locus. **A**: The unrearranged IgL locus, composed of single functional V, J, and C elements, and upstream pseudo-V genes. Triangles represent recombination signal sequences. The unrearranged IgL locus does not undergo gene conversion events. **B**: Rearrangement of the IgL locus targets the V segment for intrachromosomal gene conversion. Rearrangement occurs prior to migration to the bursa of Fabricius. During B cell development within the bursa, the rearranged V_{L1} gene is somatically diversified by sequence substitutions from the homologous pseudo V genes. This gene conversion process is nonreciprocal and occurs exclusively in *cis*. The leader (L) and J gene segments do not undergo this somatic diversification process.

of gene conversion events (3′ end defined in relationship to the transcriptional orientation of the functional V gene segment) occurred at positions of nonidentity between the donor and the recipient. In addition, 3′ ends occasionally extend into the nonhomologous J gene segment of the rearranged V–J gene segment, and are found to be resolved occasionally in an imprecise manner, leading to the duplication or loss of codons at the 3′ site of resolution of the gene conversion tract.

Finally, Thompson was able to demonstrate that the gene conversion process is specific to the rearranged immunoglobulin gene locus, and is not observed at other loci which contain transcriptionally active genes and flanking pseudogenes (Thompson, 1989). Developing B cells in the bursa of Fabricius do not undergo sequence modification of their Cα5 tubulin gene or histone genes as a result of gene conversion, despite the fact that both genetic loci contain flanking genetic segments which could be potential donors of polymorphic genetic information. This finding suggests that the immunoglobulin gene conversion that is observed during B cell development is regulated in a manner that restricts the process to the rearranged V gene segment.

The observation that both immunoglobulin recombination and gene conversion occur exclusively in cells of the B lymphoid lineage suggested that these two processes are related. Existing evidence suggested that both events might be targeted by strand cleavage at conserved heptamer recognition sequences. These heptamer sequences are components of the recombination signal sequence necessary for rearrangement. Sequences related to the recombination signal sequence heptamer are also present at conserved locations within the functional avian V gene segments of both chickens and ducks (McCormack et al., 1989c; Reynaud et al., 1989). Proteins recognizing these sequences might be involved in both the rearrangement and gene conversion processes. To investigate this possibility, Carlson et al. (1991) examined chicken lymphoid cells for the expression of the newly characterized recombinase activating genes 1 and 2 (RAG-1 and RAG-2). In mammalian species, RAG-1 and RAG-2 are both necessary and sufficient to direct V(D)J recombination in nonlymphoid cells (Schatz et al., 1989; Oettinger et al., 1990). As expected from studies in mammalian cells, cells that are undergoing recombination express both the chicken RAG-1 and RAG-2 genes. In contrast, B cells in the bursa of Fabricius, which have ceased rearrangement but are actively undergoing Ig gene conversion, express RAG-2 mRNA, but not RAG-1 mRNA (Carlson et al., 1991). The absence of the coexpression of RAG-1 and RAG-2 within the bursa of Fabricius is consistent with the fact that these cells are no longer undergoing immunoglobulin recombination. It is possible that the continued expression of RAG-2 mRNA in developing bursal lymphocytes is merely a marker of the stage of B cell development, and that the termination of V(D)J recombination is the result of the loss of RAG-1 gene expression. However, RAG-1 and RAG-2 expression are coordinately terminated in thymocytes, which use these proteins to rearrange their T cell receptor genes (Turka et al., 1991). Therefore, the differential regulation and selective maintenance of RAG-2 gene expression as B cells undergo gene conversion in the bursa of Fabricius appears to represent a specialized event which occurs only in bursal B cells. Further investigations to determine whether RAG-2 has a specific role in regulating immunoglobulin gene conversion will be necessary to resolve this issue.

ONLY CELLS THAT HAVE UNDERGONE IMMUNOGLOBULIN DIVERSIFICATION BY INTRACHROMOSOMAL GENE CONVERSION MIGRATE FROM THE BURSA OF FABRICIUS

One of the remaining questions that needs to be resolved in understanding B cell development in avian species is how the cell determines when it has completed the gene conversion process and how it leaves the bursa of Fabricius to take up residence in secondary lymphoid organs to provide the animal with a stable host defense system. One potential explanation for how this occurs was suggested by the observation that shortly after birth, an increasing number of bursal lymphocytes cease exponential growth and enter a quiescent G0-G1 phase of the cell cycle (Thompson et al., 1987). Analysis of the diversification of the rearranged V gene segments of these cells indicates that they have already undergone extensive sequence modification of their V gene segments, suggesting that these cells have completed the diversification of their immunoglobulin genes (Thompson and Neiman, 1987). Furthermore, these are the cells that appear to migrate from the bursa of Fabricius to the secondary lymphoid organs (Thompson and Neiman, 1987; McCormack et al., 1989a). Thus, these data suggest a model in which the process of gene conversion and B cell development within the bursa of Fabricius is terminated by the inability of the immunoglobulin molecule expressed on the surface of the cells to undergo further stimulation within the environment of the bursa of Fabricius. Thus, these cells cease immunoglobulin receptor-mediated proliferation, acquire a quiescent phenotype, and migrate to the periphery to take up residence as diversified humoral immune defense cells.

A MODEL FOR AVIAN B CELL DEVELOPMENT

Taken together, recent studies suggest a model of B cell development in avian species which is shown in Figure 9. The B cell lineage is initiated by the commitment of a few progenitor cells during early hematopoiesis, between days 10 and 15 of embryogenesis. Although the factors influencing

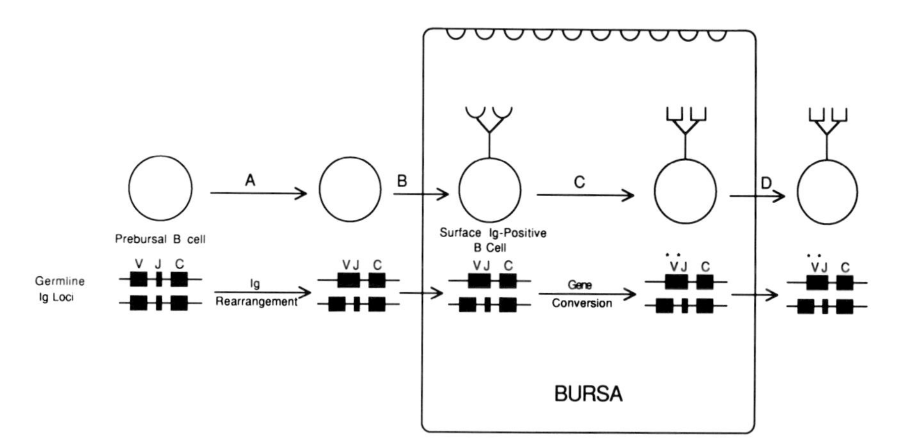

Fig. 9. A model of B cell development in avian species. Prebursal stem cells present in the splenic anlage rearrange their immunoglobulin genes and migrate to the bursa of Fabricius. Once in the bursa, Ig-expressing cells proliferate in response to antigen stimulation. The immunoglobulins expressed in these B cells are of a single specificity, that of the single functional germline-encoded V and J gene segments, and may recognize a self-antigen present in the bursa. Within the bursa, proliferating B cells undergo gene conversion of their rearranged V genes, thus altering their antigenic specificities and creating diversity within the pool of B cells. B cells which have lost specificity for the bursal self-antigen or any other self-antigen expressed in the bursa then migrate to the periphery and take up their role of producing antibody in response to foreign antigens that the bird encounters during its lifetime.

this molecular decision remain completely unknown, these committed progenitor cells acquire the ability to undergo V(D)J recombination, thus creating the antigen binding exon of the immunoglobulin heavy and light chain genes, and activating their transcription. These B cells with rearranged Ig loci also acquire the ability to migrate to the bursa of Fabricius. Within the bursa of Fabricius, B cells that have an immunoglobulin molecule composed of the unique germline V and J gene segments on their surface are induced to enter a proliferative stage of B cell development. The resulting expanded B cell pool will constitute the humoral immune system of the bird. This proliferative expansion is specific for cells that have produced functional immunoglobulin molecules composed of germline-encoded V and J segments. Therefore, the germline elements, when recombined to form an antigen-binding domain, may recognize a self-antigen expressed within the bursa of Fabricius. Thus, B cell expansion in the bursa results from an antigen-dependent proliferation. As B cells proliferate in the bursa of Fabricius as a result of exposure to antigen, they undergo Ig sequence diversification by intrachromosomal gene conversion from flanking φV gene segments on both the heavy and light chain locus. The gene conversion process results in progressive modification of the immunoglobulin molecule expressed on the B cell surface. When B cells develop to the point that their surface Ig molecules no longer recognize either the bursal-specific antigen to which they were originally directed, or any other self-antigens expressed within the bursa of Fabricius, proliferation ceases. These B cells with somatically altered Ig molecules on their surfaces then acquire the ability to migrate to the periphery, where they take up residence as part of the humoral immune defense system. By the process of intrachromosomal gene conversion, these cells will have a widely diversified immunoglobulin receptor and will have been selected for non–self-reactivity by the process of B cell development in the bursa of Fabricius. Although by no means proven, this model is testable. Thus, avian B cell development provides an excellent system for studying lineage development in a complex multicellular organism, as well as providing a model system for understanding the regulation of recombination and gene conversion.

THE STUDIES OF AVIAN B CELL DEVELOPMENT CONFIRM THAT RECOMBINATION IS USED TO REGULATE DEVELOPMENT IN A VARIETY OF SPECIES

In prokaryotic organisms, recombination has been frequently observed to activate gene transcription (Borst and Greaves, 1987), and chickens use recombination to activate transcription of the germline-encoded immunoglobulin molecule in a B cell-specific fashion. The intrachromosomal gene conversion process that creates the immunoglobulin repertoire of avian species has a number of homologies with the mating type class switching used in yeast to create diversity within a population of mating types for sexual conjugation (reviewed in Strathern, 1988). In this system, the active mat locus is targeted for gene conversion by the apparent transcription of the locus, in a manner similar to targeting of the immunoglobulin gene locus. Thus, by studying avian B cell development, we can see that two distinct strategies of molecular recombination have been utilized in series, in order to both create a lineage of cells to be used for the humoral immune system and diversify their repertoire of antigen recognition.

REFERENCES

Alt FW, Baltimore D (1982): Joining of immunoglobulin heavy chain gene segments: Implications from a chromosome with evidence of three D–J_H fusions. Proc Natl Acad Sci USA 79:4118–4122.

Borst P. Greaves DR (1987): Programmed gene rearrangements altering gene expression. Science 235: 658–667.

Carlson LC, McCormack WT, Postema CE, Barth CF, Humphries EH, Thompson CB. (1990): Templated insertions in the rearranged chicken IgL V gene segment arise by intrachromosomal gene conversion. Genes Dev 4:536–547.

Carlson LM, Oettinger MA, Schatz DG, Masteller EL, Hurley EA, McCormack WT, Baltimore D, Thompson CB (1991): Selective expression of RAG-2 in chicken B cells undergoing immunoglobulin gene conversion. Cell 64:201–208.

Cooper MD, Cain WA, Van Alten PJ, Good RA (1969): Development and function of the immunoglobulin producing system. I. Effect of bursectomy at different stages of development on germinal centers, plasma cells, immunoglobulins, and antibody production. Int Arch Allergy Appl Immunol 35:242–252.

Cooper MD, Chen CH, Bucy RP, Thompson CB (1991): Avian T cell ontogeny. Adv Immunol 50:87–117.

Eerola E, Jalkanen S, Granfors K, Toivanen A (1984): Immune capacity of the chicken bursectomized at 60H of incubation. Scand J Immunol 19:493–500.

Glick B, Chang TS, Jaap RG (1956): The bursa of Fabricius and antibody production on the domestic fowl. Poult Sci 35:224.

Glick B (1977): The bursa of Fabricius and immunoglobulin synthesis. Int Rev Cytol 48:354–402.

Granfors K, Martin C, Lassila O, Suvitaival R, Toivanen A, Toivanen P (1982): Immune capacity of the chicken bursectomized at 60 hours of incubation: Production of the immunoglobulins and specific antibodies. Clin Immunol Immunopathol 23:459–469.

Grossi CE, Lydyard PM, Cooper MD (1976): B-cell ontogeny in the chicken. Ann Immunol (Inst Pasteur) 127:931–941.

Houssaint E, Belo M, Le Douarin NM (1976): Investigations on cell lineage and tissue interactions in the developing bursa of Fabricius through interspecific chimeras. Dev Biol 53:250–264.

Huang HV, Dreyer WJ (1978): Bursectomy in ovo blocks the generation of immune diversity. J Immunol 121:1738–1747.

Jalkanen S, Granfors K, Jalkanen M, Toivanen P (1983a): Immune capacity of the chicken bursectomized at 60 hr of incubation: Surface immunoglobulin and B-L (Ia-like) antigen-bearing cells. Immunol 130:2038–2041.

Jalkanen S, Granfors K, Jalkanen M, Toivanen P (1983b): Immune capacity of the chicken bursectomized at 60 hr of incubation: Failure to produce immune, natural, and autoantibodies in spite of immunoglobulin production. Cell Immunol 80:363–373.

Lerman SP, Weidanz WP (1970): The effect of cyclophosphamide on the ontogeny of the humoral immune response in chickens. J Immunol 105:614–619.

Lydyard PM, Grossi CE, Cooper MD (1976): Ontogeny of B cells in the chicken. I. Sequential development of clonal diversity in the bursa. J Exp Med 144:79.

Mansikka A, Sandberg M, Lassila O, Toivanen P (1990): Rearrangement of immunoglobulin light chain genes in the chicken occurs prior to colonization of the embryonic bursa of Fabricius. Proc Natl Acad Sci USA 87:9416–9420.

McCormack WT, Tjoelker LW, Barth CF, Carlson LM, Petryniak B, Humphries EH, Thompson CB (1989a): Selection for B cells with productive IgL gene rearrangements occurs in the bursa of Fabricius during chicken embryonic development. Genes Dev 3:838–847.

McCormack WT, Tjoelker LW, Carlson LM, Petryniak B, Barth CF, Humphries EH, Thompson CB (1989b): Chicken IgL gene rearrangement involves deletion of a circular episome and addition of single nonrandom nucleotides to both coding segments. Cell 56:785–791.

McCormack WT, Carlson LM, Tjoelker LW, Thompson CB (1989c): Evolutionary comparison of the avian IgL locus: Combinatorial diversity plays a role in the generation of the antibody repertoire in some avian species. Int Immunol 1:332–341.

McCormack WT, Thompson CB (1990): Chicken IgL variable region gene conversions display pseudogene donor preference and 5′ to 3′ polarity. Genes Dev 4:548–558.

Moore MAS, Owen JJT (1966): Experimental studies on the development of the bursa of Fabricius. Dev Biol 14:40–51.

Oettinger MA, Schatz DG, Gorka C, Baltimore D (1990): RAG-1 and RAG-2, adjacent genes that synergistically activate VDJ recombination. Science 248:1517–1522.

Olah I, Glick B (1978): The number and size of the follicular epithelium and follicles in the bursa of Fabricius. Poult Sci 57:1445–1450.

Reynaud C-A, Anquez V, Dahan A, Weill J-C (1985): A single rearrangement event generates most of the chicken immunoglobulin light chain diversity. Cell 40:283–291.

Reynaud C-A, Anquez V, Grimal H, Weill J-C (1987): A hyperconversion mechanism generates the chicken light chain preimmune repertoire. Cell 48:379–388.

Reynaud C-A, Dahan A, Anquez V, Weill J-C (1989): Somatic hyperconversion diversifies the single V_H gene of the chicken with a high incidence in the D region. Cell 59:171–183.

Schatz DG, Oettinger MA, Baltimore D (1989): The VDJ recombination activating gene, RAG-1. Cell 59:1035–1048.

Strathern JN (1988): Control and execution of homothallic switching in *Saccharamyces cerevisiae*. In: Kucherlapati R, Smith GR (eds): "Genetic Recombintation." Washington DC: American Society for Microbiology, pp 445–464.

Thompson CB, Humphries EH, Carlson LM, Chen C-LH, Neiman PE (1987): The effect of alterations in myc gene expression on B cell development in the bursa of Fabricius. Cell 51:371–381.

Thompson CB, Neiman PE (1987): Somatic diversification of the chicken immunoglobulin light chain gene is limited to the rearranged variable gene segment. Cell 48:369–378.

Thompson CB (1989): Avian bursal lymphomas. In Melchers F, Potter M (eds): "Mechanisms of B cell neoplasia 1989." Basel: Hoffmann-La Roche Ltd., pp 46–54.

Tonegawa S (1983): Somatic generation of antibody diversity. Nature 302:575–581.

Turka LA, Schatz DG, Oettinger MA, Chun JJM, Gorka C, Lee K, McCormack WT, Thompson CB (1991): Thymocyte expression of RAG-1 and RAG-2: Termination by T cell receptor cross-linking. Science 253:778–781.

Evolutionary Conservation of Developmental Mechanisms, pages 39–53

4. Evolutionary Conservation of Developmental Mechanisms: DNA Elimination in *Drosophila*

Allan C. Spradling, Gary Karpen[1],
Robert Glaser, and Ping Zhang

Howard Hughes Medical Research Laboratory, Carnegie Institution of Washington, Baltimore, Maryland 21210

THE DOGMA OF DNA CONSTANCY

The differential gene activity theory of development holds that all somatic cells contain the same gene library but differ in which genes are actively expressed. Experiments during the last 15 years with cloned genes encoding diverse proteins have amply verified key expectations of this model. Cells do express different collections of genes, yet both active and inactive genes are detected in the genomes of all cells. These clear demonstrations have convinced many researchers that developmental changes in the genome represent an exceedingly rare and unimportant mechanism of cellular differentiation. Indeed, excepting the effects of vertebrate immunoglobulin gene rearrangements, scattered examples of gene amplification, and transposition events, somatic and germcell genomes are now widely assumed to be identical nucleotide for nucleotide.

Widespread faith in the dogma of DNA constancy has remained unshaken despite the existence of exceptional organisms that have long been known to extensively alter their germline genomes during somatic cell development (reviewed in Hennig, 1986; Pimpinelli and Goday, 1989). The alterations in these organisms appeared to fall into distinct classes that lacked any unifying mechanism or function. Cytological studies revealed that germ cell chromosomes in nematodes such as *Parascaris equorum* fragment into multiple smaller chromosomes while losing terminal heterochromatin (Boveri, 1887). In contrast, lower Dipterans such as *Miastor metrolas* expel entire chromosomes from somatic lineages (Kahle, 1908; White, 1948), while copepods such as *Cyclops strenuus* eliminate interstial heterochromatic regions without altering chromosome continuity (Beermann, 1959). Polytene cells in Dipterans such as *Drosophila* also greatly reduce their centromeric heterochromatin relative to germline or diploid cells (Heitz, 1934). However, sequence underrepresentation in polytene cells was ascribed to differences in the replication of the affected regions, rather than to elimination (Rudkin,

[1]Curent address: MBVL, The Salk Institute, 10010 N. Torrey Pines Rd. La Jolla, California 92037

1969; reviewed in Spradling and Orr-Weaver, 1987). The diverse nature of these alterations, and the fact that the eliminated or underrepresented material derives primarily from heterochromatic regions containing few if any genes, reinforced the assumption that these modifications lacked functional significance, and posed little challenge to the general validity of DNA constancy.

Studies carried out at the molecular level have started to reveal a greater similarity among these processes of genome modification than previously suspected. Ciliated protozoans, whose somatic macronucleus is produced from the germline micronucleus by a complex process involving extensive genomic changes, have been particularly informative (reviewed by Yao, 1989). During macronuclear development, thousands of specific internal DNA segments, including most repetitive DNAs, are eliminated and the chromosomes rejoined. Subsequently, hundreds or thousands of breaks occur at specific chromosome sites followed by the addition of telomere repeats to produce new chromosomes much smaller in size. Chromosome breakage and telomere addition has recently been documented in *Ascaris* (Müller et al., 1991). Furthermore, internal DNA elimination events were shown to delete a gene encoding a putative ribosomal protein from somatic cells of this species (Etter et al., 1991). Recent studies in our laboratory (reviewed below), have suggested that the underrepresentation of heterochromatic sequences in *Drosophila* polytene cells results from DNA elimination rather than underreplication (Karpen and Spradling, 1990; R. Glaser, B. Karpen, A. Spradling, unpublished). These new insights suggest that DNA elimination may in fact be an evolutionarily conserved developmental mechanism that has not been widely detected because it acts primarily on little-studied, repetitive DNA sequences.

Dp1187: A MODEL CHROMOSOME

Developmental changes in centromeric heterochromatin have been difficult to analyze for technical reasons. Centromeric heterochromatin lacks single-copy sequences that could serve as probes to study changes within a specific region during cell differentiation. Furthermore, clones derived from heterochromatic DNA are frequently unstable during propagation in *E. coli* (Lohe and Brutlag, 1986). These difficulties have prevented the mapping of any large heterochromatic region, including its junctions with euchromatin. We circumvented these difficulties by studying a *Drosophila* minichromosome called *Dp1187* (Karpen and Spradling, 1990). *Dp1187* is the smallest known functional chromosome in a multicellular organism; it consists mostly of a 1000-kb segment of centromeric heterochromatin derived from the base of the X chromosome. This region was joined artificially, by X-ray treatment, to a 290-kb segment from the normal tip of the X chromosome. It contains eight genes, including the body and bristle color gene *yellow* (see Lindsley and Zimm, 1992). The junction of these two segments is called the "breakpoint" and is denoted as position 0 on the standard *Dp1187* molecular map (Fig. 1).

The polytene chromosomes observed in larval salivary gland cells dramatically reveal the process of underrepresentation. The normal X chromosome is heterochromatic over nearly 50% of its length in diploid metaphases. However, in the salivary gland, this heterochromatin has been reduced to a tiny fraction of the total chromosome length, and becomes associated with heterochromatin from the other chromosomes in a central mass called the chromocenter (Fig.2A) (see Agard and Sedat, 1983). Other unexpected connections in which chromosome regions are joined by "ectopic fibers" are

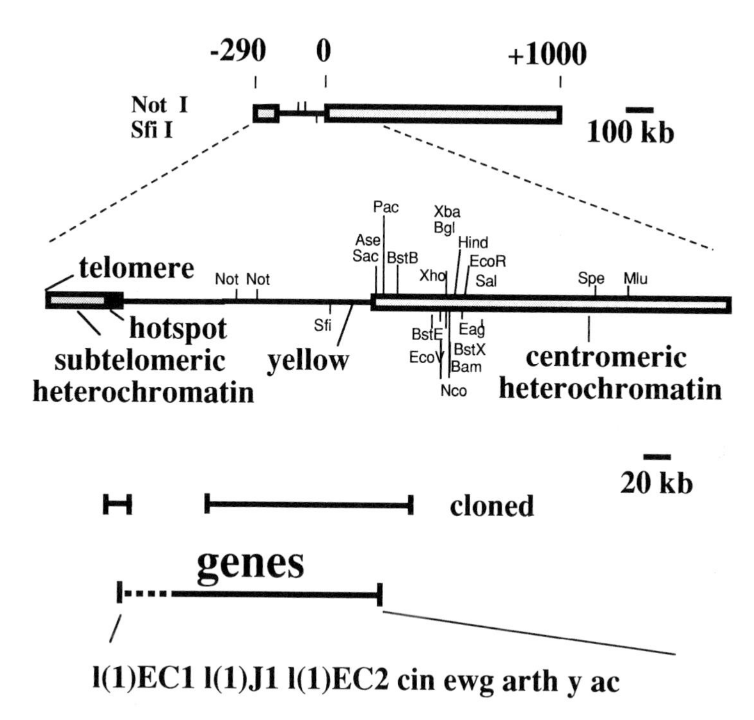

Fig. 1. Molecular structure of *Dp1187*. A restriction map of the minichromosome *Dp1187* is shown, with a coordinate system in kilobase pairs. Heterochromatin is boxed, while euchromatic regions are thick lines. The region between coordinates −290 (called the "euchromatic telomere", or "distal telomere") and +300 is expanded below. The position of specific features mentioned in the text such as the P element insertion hotspot ("hotspot"), the yellow gene ("yellow"), and the distal telomere ("telomere") are shown. Restriction sites for 21 restriction enzymes are indicated. While all the Not I and Sfi I sites are listed, only the first site encountered within the centromeric heterochromatin is listed for the other enzymes. At the bottom of the figure are shown the regions that have been cloned. The order and approximate location of the eight genes located on *Dp1187* are indicated (see also Lindsley and Zimm, 1992).

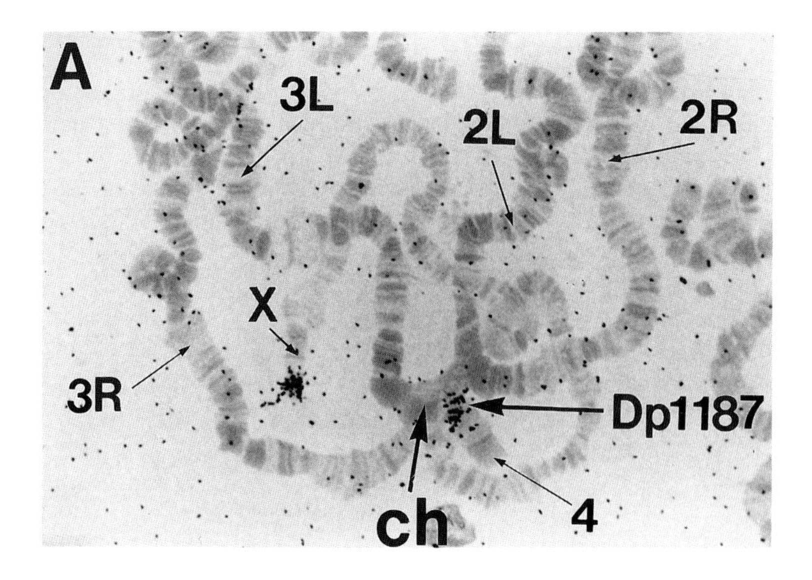

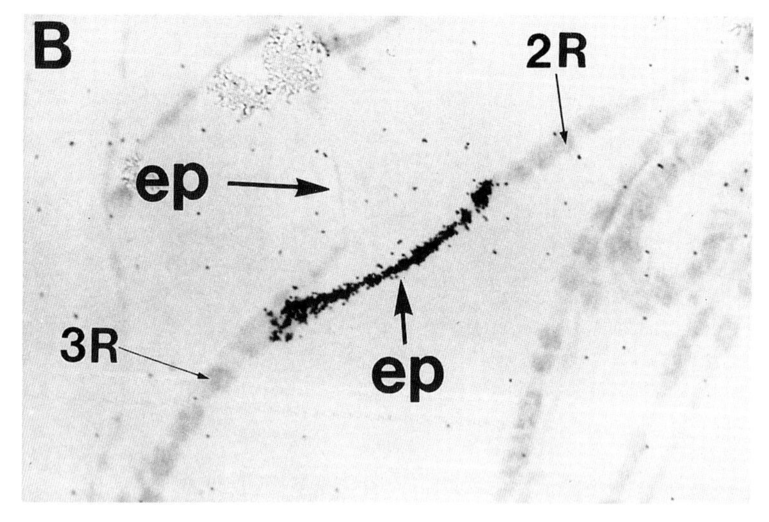

Fig. 2. Polytene chromosome structure and *Dp1187*. Squash preparations of larval salivary gland polytene chromosomes are shown. **A**: The chromocenter (ch) is near the bottom with six chromosome arms emanating from it. Chromosomes 2 and 3 are each joined near their middle (where their centromeric heterochromatin is located) to the chromocenter, from which the right and left arms of each chromosome can be seen to protrude (2R and 2L; 3R and 3L). Both the *X* chromosome and small *4th* chromosome contain centromeric heterochromatin at one end, so only one arm protrudes (X,4). The duplication contains mostly centromeric heterochromatin that is incorporated in the chromocenter; its location was visualized by hybridizing the preparation with a clone from position −55 kb (Fig. 1) that hybridizes to both the normal X chromosome tip and to the duplication (*Dp1187*). **B**: The tips of chromosome arms 2R and 3R are joined by an ectopic fiber (ep). Another ectopic fiber leaves the 3R tip on the left. In situ hybridization with a probe specific for a subtelomeric sequence cloned from the "hotspot" region of the minichromosome is shown to label DNA along one of these ectopic fibers, but not the other. All of the sequences complementary to the probe may have been eliminated from the unlabeled fiber.

seen more sporadically, but such fibers almost always link chromosome telomeres (Fig. 2B). The origin of these changes in chromosome structure is unknown (see Ashburner, 1980; Lamb and Laird, 1986). When present, *Dp1187* associates with the chromocenter; due to its small size, its presence can be revealed most clearly by in situ hybridization with a probe from the euchromatic region (Fig. 2A).

SINGLE-P ELEMENT INSERTIONAL MUTAGENESIS OF *Dp1187* USING NORMAL AND LOCAL HOPPING

Dp1187 is small enough to allow an incomplete restriction map to be generated using pulsed-field gel electrophoresis (Karpen and Spradling, 1990, 1992). It was still necessary to use unique sequence DNAs to probe Southern blots, and these were available only in a 120-kb walk from the breakpoint toward the euchromatic telomere (Fig. 1). Consequently, we investigated whether it would be possible to introduce additional genetic and unique sequence tags onto *Dp1187* by single-P element insertional mutagenesis (Cooley et al., 1988).

First, we mobilized a P element resident on the normal X chromosome in a strain that also contained *Dp1187* (Karpen and Spradling, 1992). Among 7,875 transpositions analyzed, 45 were new insertions onto *Dp1187*. However, molecular mapping of these insertions revealed that 39 had landed within a 5.4-kb "hotspot" located about 250 kb from the breakpoint (Fig.1, "hotspot"). The other insertions recovered were within the euchromatic region. Apparently P elements rarely transposed into centromeric heterochromatin, or if such transpositions did occur, they could not be detected by the methods used, which required that a marker gene on the P element continue to function.

In order to increase our collection of useful *Dp1187* derivatives, we next investigated whether the P element exhibited a special property we call "local jumping." The maize Ac transposon was shown many years ago to transpose to sites on the same chromosome near its starting location in preference to sites on other chromosomes (Van Schaik and Brink, 1959). To determine if P elements also transposed locally at elevated rates, we remobilized a P element inserted in the hotspot region (Tower et al., 1992), as well as one inserted in the *Dp1187* euchromatin at approximately position −70 (J. Tower, G. Karpen, P. Zhang, N. Craig, A. Spradling, unpublished). The rate of new insertions onto *Dp1187* increased dramatically from approximately 1/20,000 progeny, to approximately 1/50 progeny, due to the recovery of "local jumps." These new insertions frequently fell near or within the starting P element, creating "double" P elements. However, other useful new insertions were recovered throughout the *Dp1187* euchromatin, particularly when transposition was carried out in females. Minichromosomes bearing terminal deletions close to the site of the starting P element also were recovered at high frequency. However, insertions in centromeric heterochromatin still were not recovered. Figure 3 summarizes some of the useful *Dp1187* derivatives produced in these experiments.

Dp1187 UNDERREPRESENTATION IN THE SALIVARY GLAND AND OVARY

We used our knowledge of the *Dp1187* restriction map, and the existence of duplication-specific sequences present on the P insertions to analyze how sequences become underrepresented during the growth of endopolyploid cells. Three different cell types were investigated. Larval salivary gland cells are the subject of much previous work on underrepresentation, and exhibit

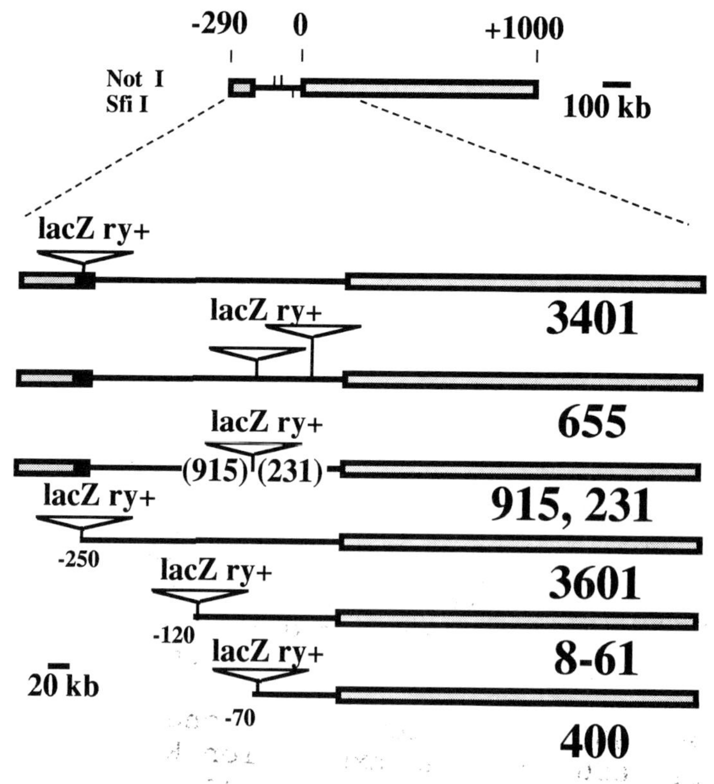

Fig. 3. Structure of *Dp1187* derivatives generated by P element insertion and excision. The structure of several derivatives containing one or more insertions of a P element transposon are shown. The transposon contains the E. *coli lacZ* gene, and the eye-color marker gene *rosy*. (ry+). In several cases, mobilization of an inserted element has led to an internal deletion (915, 231) or a terminal deletion (3601, 8-61, 400) near one end of the element.

polytene chromosomes. We also studied the nurse cells and follicle cells of the adult ovary. These cells provide a more abundant source of DNA, and have been previously shown to underrepresent sequences such as satellite DNA and ribosomal DNA that are located in centromeric heterochromatin (reviewed in Spradling and Orr-Weaver, 1987).

Figure 4 shows an example of experiments that compare the copy number of specific minichromosome regions in polytene and diploid cells. Restriction fragments comprised mainly of centromeric heterochromatin near the *Dp1187* breakpoint were drastically reduced (more than 50-fold) in salivary gland DNA. Furthermore, euchromatic sequences even 50 kb or 100 kb from the breakpoint were still underrepresented (see Karpen and Spradling, 1990). Similar experiments were carried out near the telomere using minichromosomes containing insertions within the hotspot region (Karpen and Spradling, 1992). The terminal sequence of 20–40 kb also was found to be underrepresented in salivary gland DNA.

We compared the representation of *Dp1187* regions in the salivary gland to that in the adult ovary, whose DNA is derived almost entirely from endopolyploid nurse and follicle cells (R. Glaser, G. Karpen, A. Spradling, unpublished). Figure 5 summarizes the regions of *Dp1187* that were found

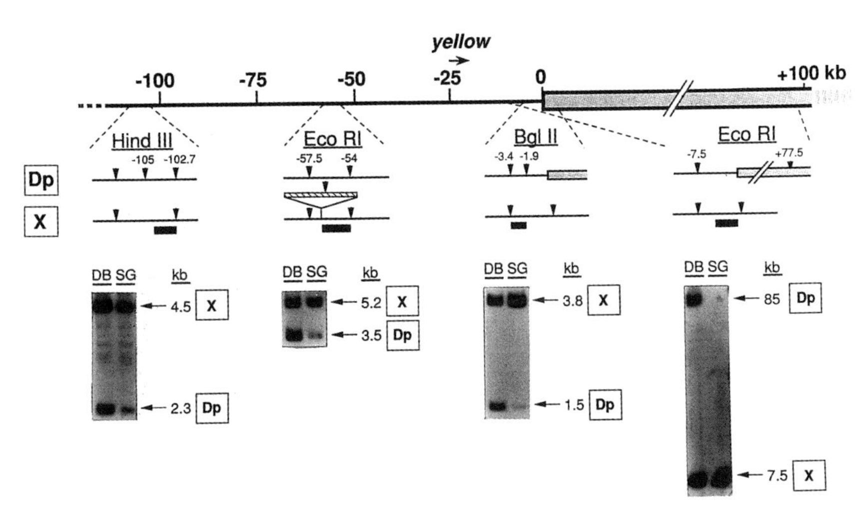

Fig. 4. DNA underrepresentation of *Dp1187* regions in salivary gland DNA. Four different DNAs were used as probes to compare the copy number of specific minichromosome regions to the corresponding sections of the normal X chromosome. Comparisons were made in DNA from both diploid imaginal disc and brains, and polytene salivary gland of *X/O* males containing one copy of *Dp1187*. Maps of the expected bands on the map and their positions of migration shown on the corresponding Southern blots. The large heterochromatic band characteristic of *Dp1187* near the breakpoint is reduced at least 60-fold relative to the corresponding euchromatic region from the *X* chromosome. Euchromatic bands were also reduced by about 40-fold at −2 kb, 8-fold at −55 kb, and 2-fold at −102 kb. (From Karpen and Spradling, 1990).

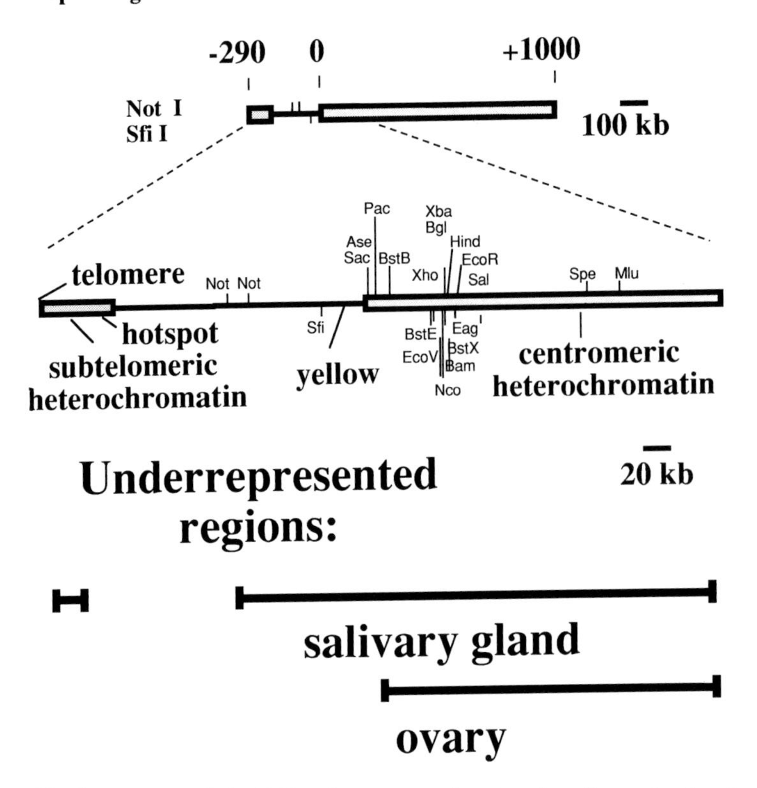

Fig. 5. Summary of *Dp1187* underrepresented regions in salivary gland and ovary. Below the map of the distal portion of *Dp1187* are shown those regions that were found to be underrepresented in DNA from larval salivary glands, or whole adult ovaries. Underrepresentation was determined by Southern blotting as shown in Figure 4. In some cases comparisons were made on pulsed-field gels where changes in both band intensity and length could be used to assess underrepresentation.

to be underrepresented in these tissues. *Dp1187* centromeric heterochromatin was extensively underrepresented in adult ovary DNA. However, unlike the salivary gland cells, nurse cells and follicle cells did not show significant changes in the copy number of the euchromatic and telomeric regions of the minichromosome.

UNDERREPRESENTATION IS CAUSED BY DNA ELIMINATION

Two models have been proposed to account for the underrepresentation of *Dp1187* regions in endopolyploid cells. The underrepresented regions could fail to replicate fully, a mechanism that has long been supposed to explain these observations. Alternatively, these regions could contain sites of DNA elimination (Karpen and Spradling, 1990). The two models predict that different structures will be present in DNA isolated from underrepresented regions. If underreplication causes the sequence reductions, then stalled replication forks should be present, as predicted previously (Laird, 1973). Alternatively, if DNA elimination occurs, with or without chromosome breakage and telomere addition, then the total size of the minichromosome should decrease in polytene tissues, and new DNA junctions should appear.

We have carried out several tests of these predictions using *Dp1187* and its derivatives (R. Glaser, G. Karpen, A. Spradling, unpublished). Two-dimensional gels that can resolve replication forks (Brewer and Fangman, 1987) were used to search for stopped forks in *Dp1187* DNA from both salivary glands and ovaries. None were found, even at levels 10 times lower than predicted by the underreplication model. The same ovary DNA preparations that lacked forks near the minichromosome breakpoint contained the expected number of forks near the amplifying chorion genes (see Heck and Spradling, 1990).

In contrast, several predictions of the DNA elimination model were verified by these experiments. Pulsed-field gel electrophoresis of uncut DNA from ovaries containing the 3401 minichromosome (Fig. 3), when probed with a *lacZ* probe specific for the P element inserted on the minichromosome, revealed bands of approximately 680 kb and 380 kb, instead of the expected 1300-kb band seen in imaginal disc or embryonic DNA (Fig. 6). Thus the minichromosome decreased drastically in size in nurse cells and follicle cells; the presence of two new bands may indicate that the changes were tissue-specific, with one band characteristic of nurse cells and the other of follicle cells. These changes implied that new junctions must have formed, at least within the centromeric heterochromatin, during the events leading to the smaller chromosomes. When large restriction fragments extending into the centromeric heterochromatin were examined on Southern blots, they were found to be reduced in intensity in ovary DNA. Heterogeneous new bands of mostly smaller size were labelled as well, suggesting that new junctions had formed but were heterogeneous in structure (R. Glaser, G. Karpen, A. Spradling, unpublished).

Previously we reported that the underrepresentation of specific *Dp1187* euchromatic regions varied between individual cells within a single salivary gland (Karpen and Spradling, 1990). This was interpreted as evidence that DNA elimination events within individual cells differ in the exact location of their breakpoints and the amount of material removed, particularly in the vicinity of the breakpoint region. Heterogeneity may be exaggerated near abnormal junctions between centromeric heterochromatin and euchromatin, since *cis*-regulatory sequences that are presumed to control elimination may no longer be able to function normally. Heterogeneous elimination was paralleled by position-effect variegation for the *yellow* gene, supporting a role for DNA elimination in this phenomenon (Spradling and Karpen, 1990, 1992; R. Glaser, G. Karpen, A. Spradling, unpublished).

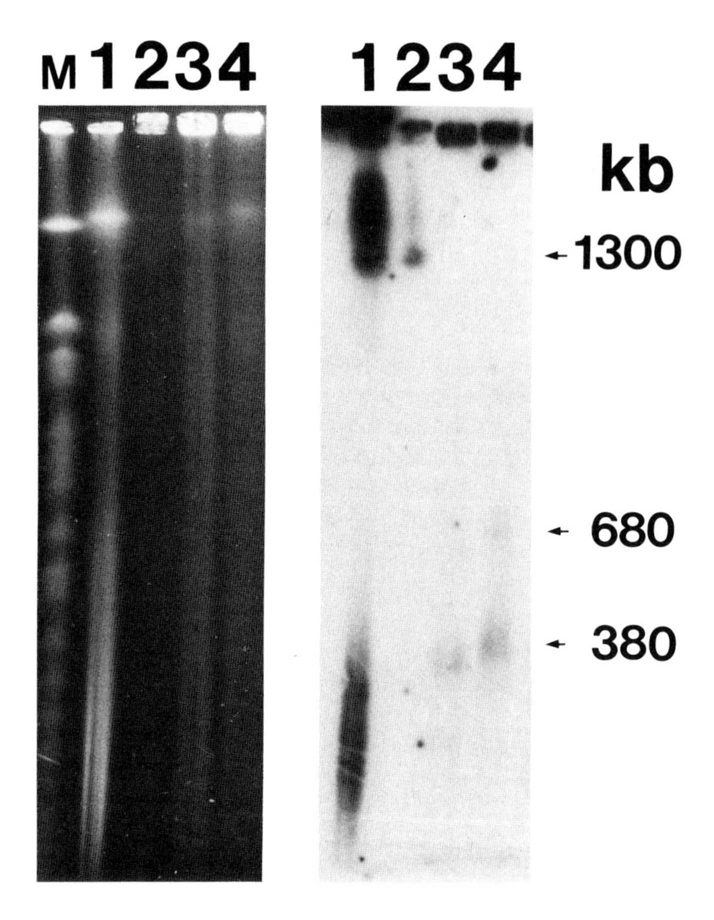

Fig. 6. *Dp1187* size in DNA from different tissues. DNA was isolated from several tissues of a strain containing *Dp3401* (see Fig. 3), and separated on a pulsed-field gel. **Left panel**: Gel stained with ethidium bromide to reveal total DNA. **Right panel**: Southern blot hybridized with an E. *coli lacZ* probe that is specific for the minichromsome. Lane M shows the migration of yeast chromosomal DNA as size markers. Lane 1: DNA from 3–20-h embryos; lane 2: DNA from third instar imaginal disks; lane 3: DNA from stage 13–14 egg chambers; lane 4: DNA from stage 9–10 egg chambers. A band corresponding to the expected size of the minichromosme, 1300 kb, is seen in the embryo and imaginal disk DNA. Bands of 680 kb and 380 kb are observed in egg chamber DNA that derives from endopolyploid ovarian nurse cells and follicle cells. The heterogeneous DNA derived from *Dp3401* in the embryo DNA may arise from endoploid cells that have already begun to form.

ROLE OF ELIMINATION IN POLYTENE CHROMOSOME STRUCTURE

DNA elimination provides a simple explanation for many of the perplexing structural properties of polytene chromosomes. Our finding that sequences near the telomere of *Dp1187* undergo elimination suggests a model for ectopic fiber formation (Fig.7). An endonuclease recognizing a repetitive sequence located in the subtelomeric regions of all the chromosomes would cleave many of its target sites during the process of polytenization. Most of the time, the chromosome fibers would rejoin after loss of some intervening material. However, rarely, ligation would mistakenly occur to a subtelomeric fiber from a different chromosome. Following further rounds of replication, this junction would produce a covalent interconnection between the two chromosome tips that upon chromosome preparation would form an ectopic fiber. Differences in elimination between the different strands could also

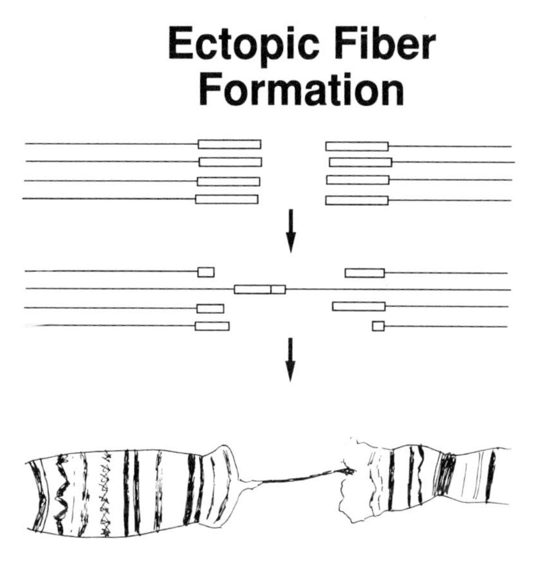

Fig. 7. Model for ectopic fiber formation. The subtelomeric regions of two separate polytene chromosome arms at an intermediate stage of cell growth are drawn schematically at the top. The four lines indicate separate DNA molecules that make up the polytene chromosomes at this stage, while the boxes indicate tandemly repeated subtelomeric DNA that is a target for elimination. Cleavage at the target sequences and religation leads to a shortening of these sequence blocks, and an occasional ligation between the two chromosomes. Following the completion of polytenization, this leads to mature polytene chromosomes joined by an ectopic fiber.

explain the observation of both labeled and unlabeled ectopic fibers emanating from the same chromosome tip (Fig. 2B).

The same mechanism would explain why the centromeric heterochromatin of all the chromosomes becomes weakly joined together in the chromocenter (Fig. 8). Elimination of the bulk of the satellite DNA, as well as other possible target sequences, occurring simultaneously on all the chromosomes, would provide many opportunities for cross-ligation events between

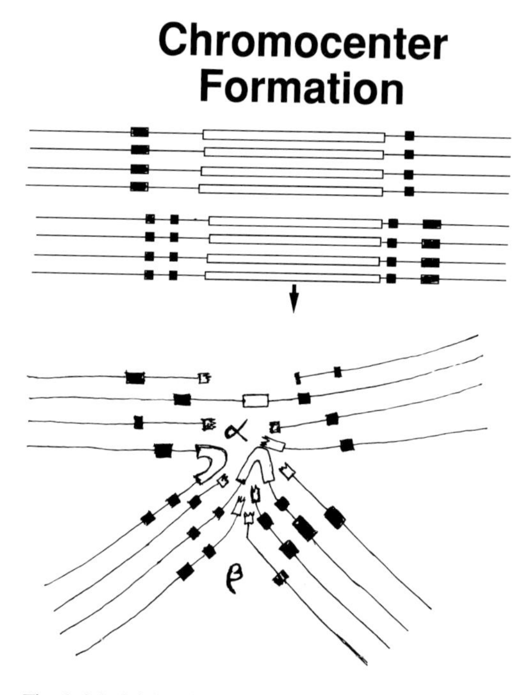

Fig. 8. Model for chromocenter formation. The centromeric regions of two separate polytene chromosomes are pictured schematically at an intermediate period of salivary gland development. Each chromosome contains four DNA strands, with various boxes depicting different repetitive sequences that are targets for DNA elimination. The large central block is postulated to include all the satellite DNAs. Following cleavage at the repeats the chromosomes are religated, but occasionally, ligation occurs to a strand of a different chromosome, leading to a polytene chromocenter with the structure indicated at the bottom. The greatly reduced satellite-containing blocks are now confined to the central portion of the chromocenter (α), while other repetitive blocks remain closer to the euchromatic arms (β). Since different amounts of elimination occured in the β region, the banding in these regions is poor.

DNA strands from different chromosomes. Even a small number of these events might create enough interconnections to hold the chromosomes together and create a chromocenter. Polytene chromosomes in some organisms lack a chromocenter; these may undergo fewer, or more precisely controlled, elimination events.

DNA elimination can explain other perplexing phenomena exhibited by insect polytene chromosomes. For example, in many midges from the family *Cecidomyiidae*, polytene chromosomes break down as development proceeds (see Ashburner, 1980), a process reminiscent of the disintegration of polytene chromosomes in the macronuclei of the ciliate *Euplotes* that occurs when elimination reaches the chromosome breakage phase (Ammerman, 1971; Tausta and Klobutcher, 1990). In other Diptera, heterochromatin granules have been reported to bud from centric regions during the growth of some polytene cells (Keyl and Hägele, 1966; Samols and Swift, 1979). These observations have long posed difficulties for the underreplication model.

DNA ELIMINATION: A CONSERVED AND FUNCTIONALLY IMPORTANT DEVELOPMENTAL PROCESS?

Our experiments suggest that DNA elimination rather than underreplication is responsible for the losses of heterochromatic chromosome regions that are observed in *Drosophila* endopolyploid cells. Since direct evidence for underreplication is lacking in the case of many other organisms where sequence reductions have been documented, including many invertebrate and plant taxa (see Nagl, 1978), DNA elimination may occur much more widely than previously supposed. Furthermore, it is likely that many examples of elimination remain to be discovered. Clones from an eliminated region would probably be required to detect this process in organisms where the total amount of elimination is less than 10–20% of the total genome. Repeated sequences further complicate detection of elimination, unless the majority of the sequences in a given family are lost. Experimental data concerning the constancy of heterochromatic regions in most organisms (including vertebrates) is extremely limited. The experiments of the late forties and fifties measured only whole genomic DNA and would not have detected changes involving even 10% of the entire genome (reviewed in Davidson, 1976).

Perhaps the principal reason why most researchers have shown little interest in DNA elimination is its lack of known function. Since in most cases, cellular genes do not appear to be lost, the phenomenon has been relegated to the status of an evolutionary curiosity, perhaps a nettlesome manifestation of "parasitic" DNAs that are able to excise themselves so as to prevent any deleterious effects on somatic cell activities. Although many possible functions have been suggested for DNA elimination (reviewed by Tobler, 1986), there has been a tendency to focus on potential functions for the eliminated material in the germline. This overlooks the possible importance of the elimination process itself, rather than the material removed. Elimination changes the structure of somatic genomes and may do so in a manner that varies between different cell types. The modifications might modulate gene function by controlling the general organization of DNA within the nucleus. This seems particularly likely with regard to genes such as ribosomal genes, that are frequently located in heterochromatic regions containing repetitive sequences. Obviously much additional support is needed before DNA elimination can be accepted as an evolutionarily conserved process of gene modulation. However, the tools we have developed in analyzing the behavior of *Dp1187* during *Drosophila* development should greatly assist in supporting or contradicting this fascinating possibility.

REFERENCES

Agard DA, Sedat JW (1983): Three-dimensional architecture of a polytene nucleus. Nature 302:676–681.

Ammermann D (1971): Morphology and development of the macronuclei of the ciliates *Stylonychia mytilus* and *Euplotes aediculatus*. Chromosoma 33:209–238.

Ashburner M (1980): Some aspects of the structure and function of the polytene chromosomes of the Diptera. In Insect Cytogenetics, Symp R Soc London 10:65–84.

Beermann S (1959): Chromatin-Diminution bei Copepoden. Chromosoma 10:504–514.

Boveri T (1887): Über Differenzierung der Zellkerne während der Furchuung des Eies von Ascaris megalocephala. Anat Anz 2:688–693.

Brewer BJ, Fangman WL (1987): The localization of replication origins on ARS plasmids in *S. cerevisiae*. Cell 51:463–471.

Cooley L, Kelley R, Spradling AC (1988): Insertional mutagenesis of the *Drosophila* genome with single P elements. Science 239:1121–1128.

Davidson E (1976): "Gene Activity in Early Development." New York: Academic Press, 452 pp.

Etter A, Aboutanos M, Tobler H, Müller F (1991): Eliminated chromatin of *Ascaris* contains a gene that encodes a putative ribosomal protein. Proc Natl. Acad Sci USA 88:1593–1596.

Heck M, Spradling A (1990): Multiple replication origins are used during *Drosophila* chorion gene amplification. J Cell Biol 110:903–914.

Heitz E (1934): Über a- and b-Heterochromatin Sowie Konstanz und Bau der Chromomeren bei *Drosophila*. Biol Zentralbl 54:588–609.

Hennig W (ed) (1986): "Germ-Line Soma Differentiation." Berlin: Springer-Verlag, 196 pp.

Kahle W (1908): Die Paedogenesis der Cecidomyiden. Zoologica 21:1–80.

Karpen GH, Spradling AC (1990): Reduced DNA polytenization of a minichromosome region undergoing position-effect variegation in *Drosophila*. Cell 63:97–107.

Karpen GH, Spradling AC (1992): Analysis of subtelomeric heterochromatin in the *Drosophila* minichromosome *Dp1187* by single P element insertional mutagenesis. Genetics (in press).

Keyl HG, Hägele K (1966): Heterochromatin-Proliferation an den Speicheldrüsen-Chromosomen von Chironomus melanotus. Chromosoma 19:223–230.

Laird CD (1973): DNA of *Drosophila* chromosomes. Ann Rev Genet 7:177–204.

Lamb MM, Laird CD (1986): Three euchromatic DNA sequences under-replicated in polytene chromosomes of *Drosophila* are localized in constrictions and ectopic fibers. Chromosoma 95:227–235.

Levis R (1989): Viable deletions of a telomere from a *Drosophila* chromosome. Cell 58:791–801.

Lindsley DL, Zimm GG (1992): "*The genome of Drosophila melanogaster.*" New York: Academic Press 1,133 pp.
Lohe AR, Brutlag DL (1986): Proc Nat Acad Sci USA 83:696–700.
Müller F, Wicky C, Spicher A, Tobler H (1991): New telomere formation after developmentally regulated chromosomal breakage during the process of chromatin diminution in *Ascaris lumbricoides*. Cell 67:815–822.
Nagl W (1978): "Polyploidy and Polyteny in Differentiation and Evolution." North-Holland/Amsterdam: Elsevier.
Pimpinelli S, Goday C (1989): Unusual kinetochores and chromatin diminution in *Parascaris*. Trends Genet 5:310–315.
Rudkin G (1969): Non-replicating DNA in *Drosophila*. Genetics 61:227–238.
Samols D, Swift H (1979): Characterization of extrachromsomal DNA in the flesh fly *Sarcophaga bullata*. Chromosoma 75:145–159.
Spradling AC, Orr-Weaver T (1987): Regulation of DNA replication during *Drosophila* development. Ann Rev Genet 21:373–403.
Spradling AX, Karpen GH (1990): Sixty years of mystery. Genetics 126:779–784.
Tausta LS, Klobutcher LA (1990): Internal eliminated sequences are removed prior to chromosome fragmentation during development in *Euplotes crassus*. Nucl Acids Res 18:854–853.
Tobler H (1986): The differentiation of germ and somatic cell lines in nematodes. In Hennig W (ed): "Germ-Line Soma Differentiation." Berlin: Springer-Verlag, pp 1–69.
Van Schaik NW, Brink RA (1959): Transposition of modulator, a component of the variegated pericarp allele in maize. Genetics 44:725–738.
Yao MC (1989): Site-specific chromsome breakage and DNA deletion in ciliates. In Berg D, Howe M (eds): "Mobile DNA." ASM Washington DC: ASM Publications, pp 715–754.
White MJD (1948): "Animal Cytology and Evolution, 3rd Edition." Cambridge, England: Cambridge University Press, 961 pp.

Evolutionary Conservation of Developmental Mechanisms, pages 55–70

5. Specification of Reproductive Cells in *Volvox*

Andrew Ransick
Division of Biology, California Institute of Technology,
Pasadena, California 91125

INTRODUCTION

Volvox is a genus of green-flagellates whose individuals possess only two cell types: biflagellate somatic cells and immotile reproductive (germ) cells. This level of organization is the simplest that a truly multicellular form can possess, and thereby makes *Volvox* development an attractive system in which to study mechanisms of cell determination. The obvious issue in studies of *Volvox* asexual development is the nature of the process that determines whether a cell remains totipotent and serves as a germ cell, or becomes committed to carrying out only somatic functions. This chapter reviews the current state of our understanding of the causal mechanisms by which the germ cells are specified during *Volvox* asexual development. Although both sexual and asexual development have been described for most *Volvox* species, for simplicity this discussion focuses only on asexual developmental patterns. Also, although there are at least eighteen recognized species in the genus *Volvox*, this discussion deals only with one group of species, collectively known as the section Merrillosphaera, which are similar developmentally and morphologically (Smith, 1944). Information will be drawn from descriptive, experimental and genetic analyses, with emphasis on my own work on *V. obversus*.

ASEXUAL DEVELOPMENT IN MERRILLOSPHAERA

The mature asexual form of all the *Volvox* species included here is a motile spheroid containing gelatinous extracellular matrix and bounded externally by a fibrous layer of matrix in which the biflagellate somatic cells are embedded; there are no cytoplasmic connections between cells (Fig. 1A). The asexual reproductive cells, called gonidia, are relatively few in number and are attached to the inner surface of the somatic layer. Asexual development in Merrillosphaera species is also characteristic: gonidia undergo a prolonged growth phase (2 to 7 days), during which they reach a large diameter (30–90μm); then they undergo 9 to 12 closely spaced rounds of cell division,

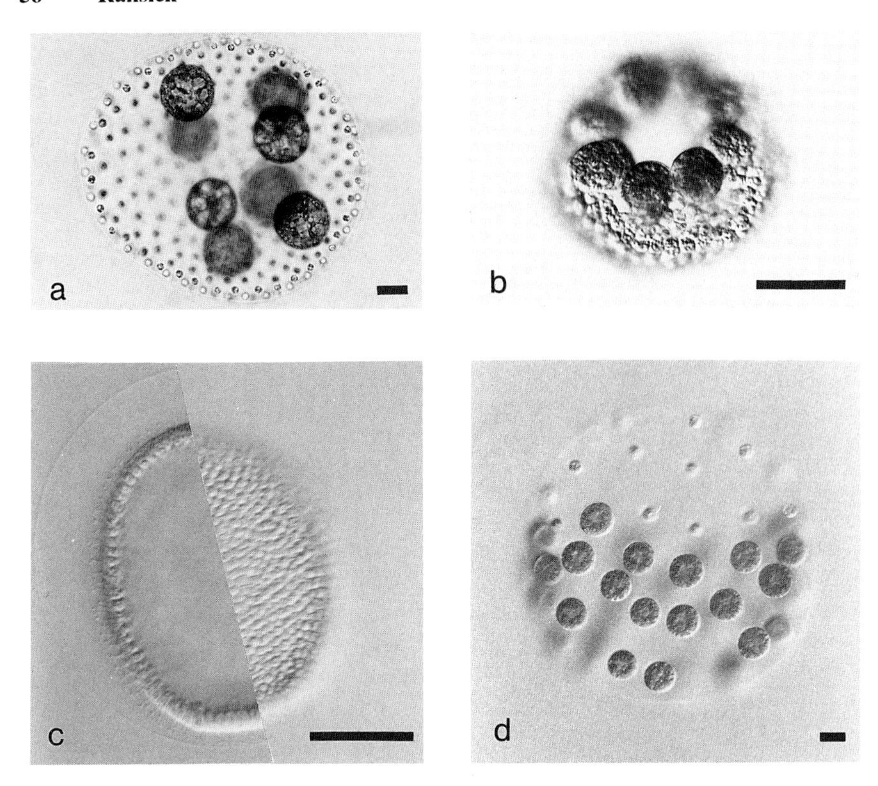

Fig. 1. (a) Asexual spheroid of *V. obversus* with mature gonidia. (b) Late cleavage stage of *V. obversus* clearly showing the eight gonidial precursors along the anterior margin of the embryo. (c) Juvenile spheroid of *V. spermatosphaera*: equatorial focus on the left, surface view of same spheroid on the right to show that all the cells are the same size. (d)Mature asexual spheroid of *Pleodorina californica*. The enlarged posterior cells will divide to form new colonies; the small cells do not divide under normal conditions. Scale bars = 30μm.

designated as embryogenesis, to generate juvenile individuals containing a new complement of somatic cells and prospective gonidia.

Table I summarizes relevant developmental characteristics of species in the section Merrillosphaera. For the purpose of this discussion, the important feature to note is the variability in the point when the prospective gonidia can first be distinguished. During asexual development in *V. carteri, V. obversus*, and *V. africanus* the prospective gonidia becoming evident at middle cleavage stages by virtue of being set aside by asymmetric cell divisions (Fig. 1B), (Starr, 1969; Karn et al., 1974). The best known pattern is that seen in *V. carteri*, forma *nagariensis*, where 16 prospective gonidia are set aside by a regular pattern of asymmetric cleavage divisions beginning at the sixth cleavage, with these larger cells differentiating directly into young gonidia following embryogenesis. A variation on the above pattern is found in asexual development of *V. gigas* and *V. tertius*, where the prospective

TABLE I.
Selected Characteristics of Asexual Development in Merrillosphaera Species

Species	All cells initially differentiate as somatic cells	Stage when prospective gonidia can be distinguished	Number of gonidia (optimal)	Approx. total cell number	Diameter of mature gonidia	References
V.powersii	YES	juvenile spheroid	35–60 (70)	500–1000	42μm	Vande Berg and Starr (1971)
V.spermatosphaera	YES	juvenile spheroid	25	1100–2600	33–36μm	Powers (1908); Smith (1944)
V.gigas	YES	inversion	25–50	1000–2000	60–70μm	Vande Berg and Starr (1971)
V.africanus	NO	late cleavage	2–4 (8)	3000–5000	50–60μm	Smith (1944)
V.tertius	NO (?)	inversion	2–8 (12)	500–2000	30–40μm	Pocock (1938)
V.obversus	NO	4th cleavage	6–9 (8)	500–1000	50–70μm	Ransick (1991)
V.carteri						
f.nagariensis	NO	6th cleavage	8–21 (16)	1000–3000	70–90μm	Starr (1969)
f.weismannia	NO	5th cleavage	8–12 (12)	2000–6000	80–100μm	Kochert (1968)
f.kawasakiensis	NO	6th cleavage(?)	8–14(30)	1000–3000	≤78μm	Nozaki (1988)
[*Pleodorina californica*]	YES	1–3 days	24–28	32	24–36μm	Goldstein (1964)
			30–42	64	30–70μm	Kikuchi (1978)
			70–90	128		

gonidia are reported to be twice the diameter of somatic precursors at the completion of the cleavage period (Vande Berg and Starr, 1971; Pocock, 1938). How these cell size differences are created has never been described, but they probably result from cells dropping out of the cleavage cycle early. Another pattern is found in *V. spermatosphaera* and *V. powersii*, where all the cells are morphologically identical to somatic cells at the completion of embryogenesis (Fig. 1C). The prospective gonidia become distinguishable many hours later as a subset of cells that transdifferentiate by retracting their flagella and enlarging differentially (Vande Berg and Starr, 1971; Powers, 1908).

Even though three different patterns for the emergence of the gonidial lineage are described above, the mature asexual spheroids of all Merrillosphaera species are organized similarly and express the same two cellular phenotypes. Therefore it is reasonable to ask whether all Merrillosphaera species use the same basic mechanisms for specifying cell types. Since the cells that mature into functional reproductive cells are generally larger at the end of embryogenesis, it would seem logical to propose that cell size plays a role in specifying reproductive cells (Pall, 1975). However, the lack of any size differences at the end of embryogenesis in some species means that in those forms some mechanism other than size must initially specify the prospective gonidial cells.

The comparative study of mechanisms of cell specification among these *Volvox* species could become more focussed if their evolutionary relationships were known. For example, if it was known which species have evolved from other *Volvox* species, they could become the focus for understanding precisely what changes at the cellular and molecular levels account for species-specific developmental traits. If, on the other hand, it was known which species, if any, have independent origins from a colonial ancestor(s), they would be useful in establishing an inventory of the genes necessary to generate the cellular dichotomy that characterizes *Volvox* organization. Even though the evolutionary relationships have not been established, it's probable that the more ancestral condition is that where all the cells first differentiate as somatic cells, and a subset then redifferentiates as the reproductive cells. This speculation follows from consideration of the developmental patterns of the closely related colonial genera, *Eudorina* and *Pleodorina*. In these forms, all the cells have a vegetative (or somatic) morphology in the young colony, but after several days a subset of cells enlarges to function as the reproductive cells (Fig. 1D) (Goldstein, 1964; Starr, 1970). The more derived (or advanced) pattern is proposed to be that where the gonidia differentiate directly from embryonic cells at the end of the cleavage period, as is seen in species with asymmetric divisions.

The mechanism of specifying gonidia has been studied in two species, *V. carteri* and *V. obversus*, both of which set aside reproductive cell precursors early in embryogenesis via asymmetric cell divisions. The experimental evidence is reviewed below, but in summary it appears that *V. obversus* localizes reproductive cell potential to a specific cytoplasmic domain and then segregates it to a subset of cells via the subsequent cleavage pattern (Ransick, 1988, 1991). In contrast, it appears that in *V. carteri*, reproductive cell potential is not localized per se, but is expressed only in cells with a relatively large volume, thus making expression of reproductive cell potential completely dependent on the occurrence and placement of asymmetric cell divisions (Kirk et al., 1991).

SPECIFICATION IN *Volvox obversus*

Embryogenesis in *V. obversus* is summarized in Figure 2. Asexual development will be summarized only briefly here since detailed descriptions are already available (Karn et al., 1974; Ransick, 1988, 1991). The most important feature to note is that polarized asymmetric divisions of the anteriormost blastomeres at the fourth through ninth cleavage cycles produce eight large cells at the anterior extremity of the embryo that differentiate as gonidia, while all other blastomeres cleave symmetrically to yield somatic cell precursors. This cleavage pattern is highly reproducible and over 98% of the embryos produce eight gonidia.

Blastomere Isolation Experiments

When blastomeres were isolated at the 2-cell through the 64-cell stages they cleaved in the normal pattern and produced the same complement and spatial distribution of cell types as they would have in the intact embryo (Table II). This demonstrates that even when cell divisions occur in isolation, reproductive cell potential segregates to the large, more anterior daughter cells starting from the fourth cleavage cycle onwards. This finding rules out any significant role for a global patterning mechanism or for cell–cell interactions in the establishment of developmental polarity and in the segregation of developmental potential.

Having established that the informational cues that influence cell fate originate from within the cells themselves, the task became finding out when these decisions concerning cell fate are made and what aspect(s) of cellular organization are most influential in the choice of fate. Four types of microsurgical manipulations were performed to address these questions: deletions of cytoplasm at various development stages, alterations of the planes of cleavage, and two different methods of altering blastomere volume.

Deletions of Cytoplasm at Various Developmental Stages

Cytoplasmic deletions were employed to reveal whether any causal link could be demonstrated between cytoplasmic organization and determinative

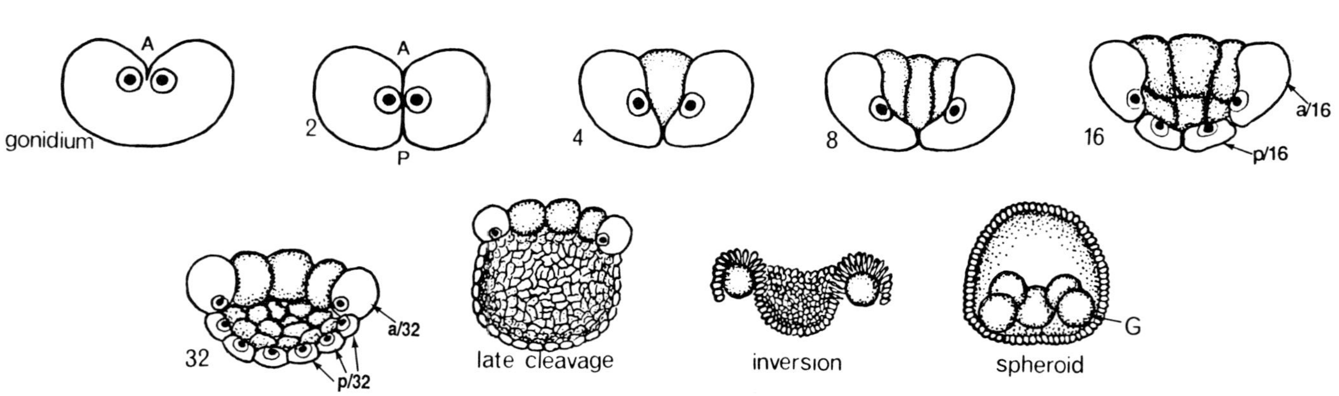

Fig. 2. Diagrammatic summary of *V. obversus* embryogenesis. Selected cross-sectional views from first cleavage through inversion illustrate the overall organization of the embryo, as well as changes in the shape of the cells and the location of the nuclei. Anterior is up in all drawings. The first cleavage furrow is initiated on the anterior side (A) of the gonidium. Completion of the first cleavage establishes the embryonic anteroposterior axis (A–P). During the 4-and 8-cell stages the embryo obtains a bowl shape, while the nuclei remain in close association with the plasma membranes lining the embryo's concave surface. The fourth cleavage is unequal and yields large anterior cells (a/16) and small posterior cells (p/16). The most anterior blastomeres continue to divide unequally, ultimately yielding eight large reproductive cells (G) in juvenile spheroids.

TABLE II.
Summary of Blastomere Isolation Experiments

Fraction of embryo isolated	Expected no. of gonidia per isolate	% Isolates with expected no. of gonidia (n)	Avg. no. of cells per isolate (n)
1/1[a]	8	100 (200)	488 (15)
1/2	4	100 (30)	232 (12)
1/4	2	100 (80)	124 (37)
1/8	1	98 (110)[b]	60 (40)
a/16	1	93 (112)[b]	31 (24)
a/32	1	98 (44)[b]	17 (14)
a/64	1	100 (39)	15 (28)
p/16	0	99 (112)[c]	27 (19)
p/32	0	100 (51)	17 (51)

[a]1/1 = intact embryo; see Figure 2 for explanation of a/16, a/32, p/16, p/32.
[b]other cases with 2 gonidia.
[c]1 case with 1 gonidium.

events. When as much as 60% of the cytoplasmic volume was deleted from uncleaved gonidia, the gonidia fragments still underwent a normal developmental program that produced juvenile spheroids containing eight gonidia, albeit of reduced size. However, deletion of 70–75% of the gonidial volume, which produced gonidia fragments 30–40 μm diameter, consistently led to developmental pattern defects—namely, the production of spheroids with less than eight gonidia. The origin of these abnormalities was traced to unequal early cleavages, which produced some disproportionately small blastomeres by the 4-and 8-cell stages (Fig. 3). Normally, the blastomeres acquire a polarized organization during the 4-and 8-cell stages, with the nucleus reaching a relatively posterior position that predisposes the 8-cell blastomeres to divide asymmetrically. In these experimental embryos, the smallest blastomeres remained apolar and cleaved symmetrically at all subsequent divisions to yield only somatic cells.

Cytoplasmic deletions were also carried out on individual blastomeres at the 4-and 8-cell stages, as well as on the anteriormost blastomeres of embryos at the 16-cell, 32-cell, and 64-cell stages, (Table III). When 10–75% of the cell volume was deleted from anterolateral regions of 4-cell blastomeres, the development of some cell fragments was defective with respect to gonidia production, while others developed normally. The loss of reproductive cell potential seemed to be independent of the amount of cytoplasm deleted, but was dependent on whether the cell fragment established a polarized organization after the deletion. When the cell fragment failed to establish a polarized organization, the subsequent cleavages were symmetrical and only somatic cells were produced; however, when a typical polarity was established, asymmetric divisions occurred and gonidia of reduced size were produced. This ability to "regulate" after having anterior cytoplasm deleted

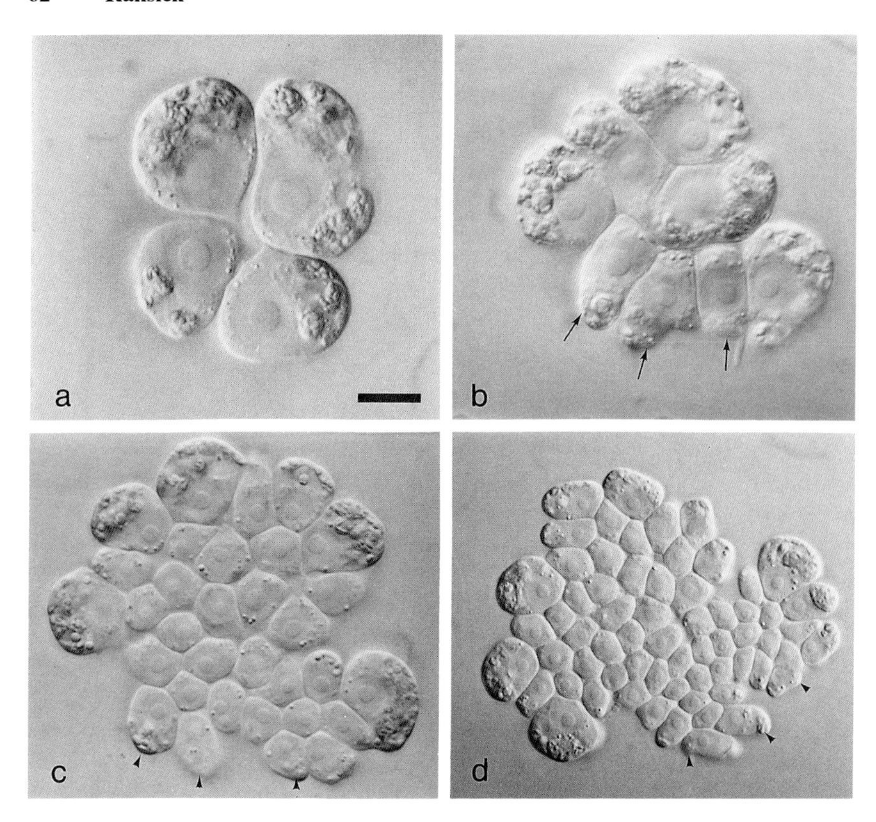

Fig. 3. Developmental sequence showing an abnormal cleavage pattern in an embryo derived from a 25% gonidium fragment (i.e. 75% deleted): (a) 4-cell stage with two cells distinctly smaller (scale bar =10μm); (b) 8-cell stage with three relatively small apolar cells (arrows); (c) 32-cell stage and (d) 64-cell stage (arrowheads point out the progeny of the cells indicated with arrows in (b). This embryo produced only five gonidia.

diminished considerably by the 8-cell stage, such that when approximately 30% of the cell volume was deleted from the anterior end of 8-cell blastomeres, they cleaved symmetrically and produced only somatic cell progeny in 85% of the cases. Similarly, when ≥16% of the anterior cytoplasm was deleted from the anteriormost blastomeres at the 16-, 32-, and 64-cell stages, 83% of all the cases cleaved symmetrically and produced only somatic cells (Table III).

These results reveal that cytoplasmic reorganization occurs during the 4-cell stage and it culminates in the establishment of a polarized organization in 8-cell blastomeres. When the anterior lobe of cytoplasm is deleted from an 8-cell stage blastomere (or later stage cell), the cell fragments cannot regenerate the polarized organization. The consequences are that the subsequent divisions are symmetrical, the cells obtain relatively small sizes, and reproductive cell potential is lost. Another important finding was that reproductive

TABLE III.
Summary of Cytoplasmic Deletion Experiments

Developmental stage	Volume deleted %	No.	Normal pattern[a] %	Pattern defects[a] %
Uncleaved Gonidium (e.g. 40% deletion)	15-40	24	100	0
	50-60	12	83	17
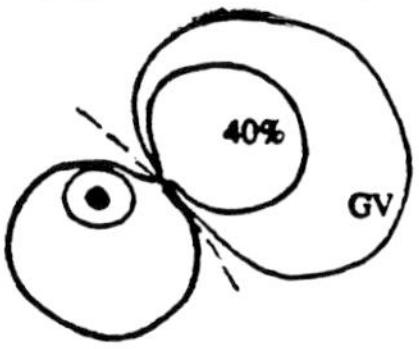	70-75	6	0	100
4-Cell (e.g. 45% deletion)	10-15	9	33	67
	20-25	15	47	53
	30-35	16	25	75
	40-45	15	47	53
	75	2	50	50
8-Cell (e.g. 30% deletion) 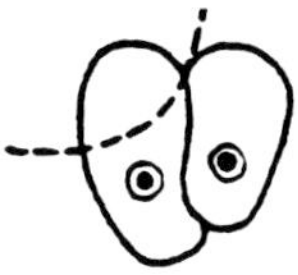	30	21	15	85
16-, 32- and 64-Cell (e.g. 40 % deletion at 32-cell)	10-15	19	37	63
	16-30	16	19	81
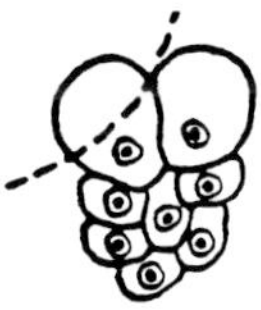	≥35	8	13	87

[a]"Normal pattern" refers to production of spheroids with eight gonidia in the case of uncleaved gonidia or 4-cell embryos, but for later stages refers to embryo fragments with two gonidia, since these deletions were carried out on cells isolated as pairs at 8-cell stage; "pattern defects" refers to cases where less than the normal number of gonidia were produced.
GV = gonidial vesicle.

cell potential could still be eliminated by cytoplasmic deletions at the 32- and 64-cell stages. This suggests that the initial rounds of asymmetric cleavage do not elicit a commitment from the anteriormost cells to produce gonidia, but do function to maintain them in a totipotent stem-cell condition.

Alteration of Cleavage Planes

As a whole, the deletion experiments reveal that by the fourth cleavage a developmental polarity is established in 8-cell blastomeres which makes the anterior lobe necessary for reproductive cell specification. But, is the role of the anterior lobe primarily to produce large cells by effecting the symmetry of the cleavages? Or is anterior cytoplasm qualitatively different from more posterior cytoplasm? If anterior cytoplasm contains some kind of "gonidial determinants," then it follows that if it were distributed to more than eight cells, more than eight cells would retain the potential to produce gonidia. As a test of this hypothesis, cleavage planes were altered by compressing embryos during the fourth cleavage. When the cleavage planes were altered such that the anterior lobe cytoplasm was partitioned into nine or ten blastomeres at the 16-cell stage, these embryos were found to have more than the usual eight asymmetrically dividing cells, and they went on to produce nine or ten gonidia (Ransick, 1991).

The compression experiments confirm that the specification of reproductive cells in *V. obversus* involves the localization of developmental potential to the anterior cytoplasm, and that the number of gonidia is dependent on the cleavage pattern subsequent to the establishment of developmental polarity. Unfortunately, since both the cytoplasmic deletions and the compression experiments cause *both* qualitative and quantitative changes in the anterior blastomeres, neither set of experiments allows a definitive conclusion about whether the major factor in specifying reproductive cells is the relatively large size of the most anterior cells or some special properties of the cytoplasm (or plasmalemma) that they inherit.

Alteration of Cell Volumes

In order to address directly the role of cell size in specifying reproductive cells, experiments were designed that altered the volume of blastomeres without also causing the type of qualitative changes in the cellular contents that are side effects of cytoplasmic deletion and compression experiments.

To test whether an increase in cell volume of posterior cells could induce them to be specified as gonidia, a method was devised to produce enlarged posterior cells. By ligating posterior gonidial cytoplasm until after the fourth cleavage and then releasing the ligation, a significant increase in volume was achieved in a posterior 16-cell stage blastomere (p/16-cell) as the extra cytoplasm merged into it (Fig. 4). The typical cleavage pattern observed after such a manipulation was that the enlarged cells continued cleaving symmetrically, dividing up the extra cytoplasm equally among the progeny such that a patch of cells with diameters ranging from 10–15 μm was present at the end of embryogenesis. These cells always differentiated into functional somatic cells. No gonidia ever differentiated from these enlarged posterior cells, even though they were five times the volume of typical somatic cells, and they were larger than some of the anteriormost cells that did differentiate as gonidia in spheroids derived from small gonidia fragments. This result demonstrates that large cell size alone cannot cause cells to be specified as gonidia in *V. obversus*. It is also consistent with the idea that posterior cytoplasm does not have the potential to specify cells as gonidia, which confirms the hypothesis that anterior and posterior cytoplasm are qualitatively different by the fourth cleavage.

In order to decrease the volume of anterior blastomeres, developing gonidia were isolated from the intraspheroidal compartment of the parental spheroid before they reached the typical mature gonidial diameter of 65–75μm. In isolation, their growth was slowed and embryogenesis was initiated at a much smaller diameter (as small as 18 μm diameter). The advantage of this technique was the generation of miniature embryos, in which the anteriormost blastomeres were even smaller than the anteriormost cells in

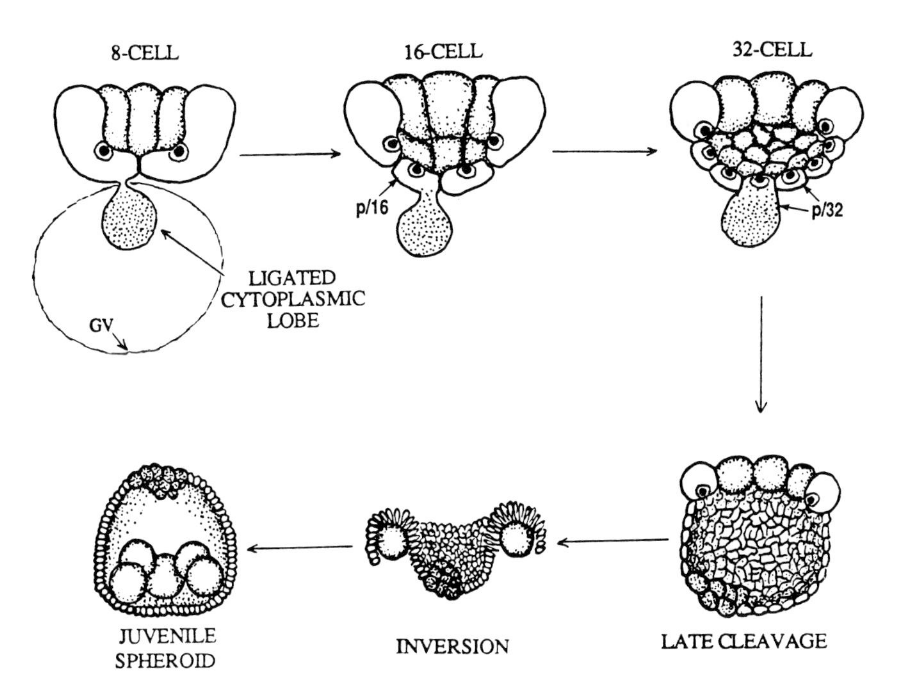

Fig. 4. Diagrammatic summary of the posterior ligation experiment described in the text. GV = gonidial vesicle.

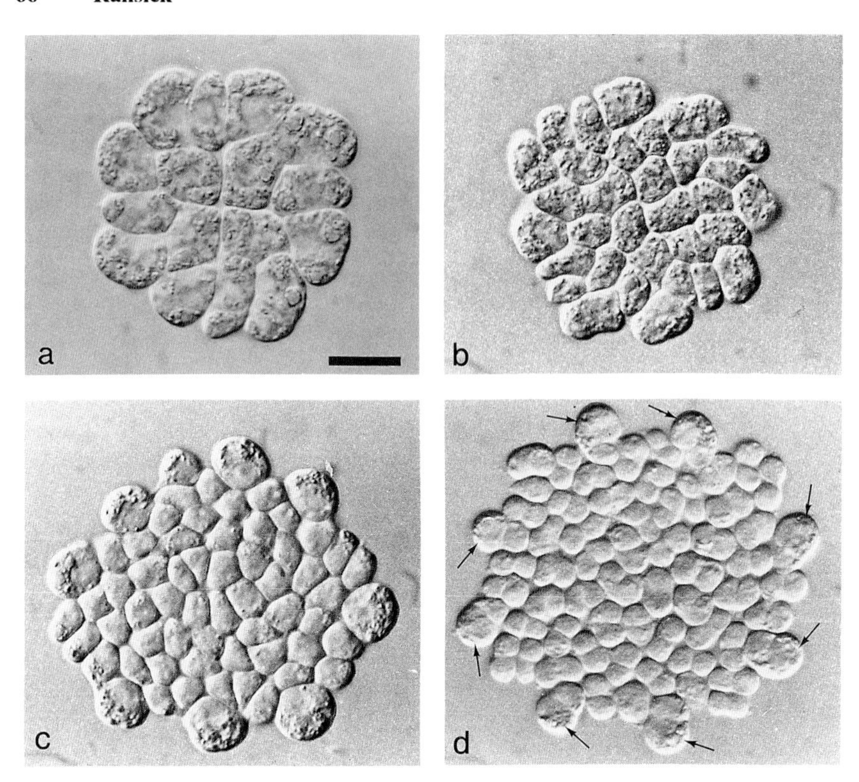

Fig. 5. Developmental sequence of 25μm diameter *V. obversus* gonidium: (a) 16-cell stage showing that the fourth cleavage was equal (scale bar = 10μm); (b) 32-cell stage with slightly larger a/32-cells; (c) 64-cell stage with eight (possibly nine) distinctly larger a/64-cells; the sixth cleavage was noticeably asymmetric, note also that all the cells are less than ten microns in diameter; (d) 64-cell embryo undergoing seventh cleavage; all the cells are dividing symmetrically except the a/64-cells (arrows). This embryo underwent inversion with 128 cells and contained nine prospective gonidia which averaged 8.5 μm diameter; however, only eight matured into functional gonidia.

embryos derived from gonidia fragments. For example, gonidia that cleaved when only 25 μm in diameter typically divided seven times to produce miniature spheroids with eight prospective gonidia that all differentiated normally—even though they averaged only 9.5 μm in diameter at the end of cleavage. The smallest cell that differentiated into a functional gonidium in these experiments was only 7 μm in diameter at the end of cleavage. The most significant finding was that in these cases all cleavages were equal until the fifth or sixth division cycle; but then slightly unequal divisions of the anteriormost cells generated a very minimal size difference (Fig. 5). This result demonstrates that relatively tiny anterior blastomeres can establish developmental polarity and segregate reproductive cell potential at the fourth and fifth cleavages even in the absence of an obvious physical asymmetry between anterior and posterior blastomeres.

It was also established, however, that cell size does ultimately have a role in maintaining the potential of anterior blastomeres to differentiate as gonidia. The limit to which the volume of the anteriormost blastomeres can be reduced without abolishing gonidial potential was discovered when isolated gonidia underwent embryogenesis at diameters of only 18–24 μm. The resulting embryos produced less than eight prospective gonidia, and some of the prospective gonidia that were produced did not differentiate into functional gonidia. When less than eight prospective gonidia were formed, the deficiency was correlated with some lineages failing to undergo asymmetric divisions. Apparently around a diameter of 25 μm is the threshold gonidial size that is required for production of a full complement of gonidia in the next generation. In embryos produced from gonidia smaller than this, some of the anterior blastomeres become so small during embryogenesis that they are completely converted over to the somatic pathway of differentiation. When prospective gonidia formed, but failed to differentiate into functional reproductive cells, they initially grew to become significantly larger than the somatic cells, but their growth then became stunted and they often differentiated an eyespot, which is an organelle required for phototaxis and normally present only in fully differentiated somatic cells. These cells of intermediate size apparently received mixed signals as to how they should differentiate, suggesting that when the size of prospective gonidia is reduced to about 7 μm diameter at the completion of embryogenesis, both the somatic and gonidial programs of differentiation become active—at least temporarily.

In summary, there is evidence that reproductive cell specification in *V. obversus* involves both quantitative and qualitative aspects of cell organization. The qualitative component becomes localized to the anterior cytoplasm and functions late in embryogenesis (or in juvenile spheroids) to activate the program of gonidial differentiation. However, activation of these "gonidial determinants" is dependent on the quantitative component, which is related to cell volume. Thus, the role of the asymmetric cleavages is clarified: in normal development the initial rounds of asymmetric divisions do not play a crucial role in specifying reproductive cells, since reproductive cell potential still segregates in 8-and 16-cell stage blastomeres of miniature embryos in which the early asymmetric cleavages had been abolished. However, asymmetric divisions must occur at some point in embryogenesis, since they are the only mechanism available to this species for maintaining cell volume above the threshold required for the process of gonidial differentiation to occur properly. The molecular nature of the qualitative and quantitative features involved in gonidial specification is unknown.

SPECIFICATION IN *Volvox carteri*

Another species in which asymmetric divisons can be clearly observed during embryogenesis is *Volvox carteri,* of which there are generally three recognized forms. Several of their distinguishing developmental characteristics are listed in Table I. Since *V. carteri,* forma *nagariensis* is the best studied of the three forms, its developmental pattern will be used as an example.

During asexual embryogenesis of *V. carteri,* f. *nagariensis,* the visible differences from *V. obversus* are:

1. The asymmetric cleavages first occur at the sixth division cycle.
2. The polarity of the asymmetric divisions is reversed, so that the larger daughter cells are more posterior.
3. All cells in the anterior half of the embryo can divide asymmetrically.
4. The asymmetric divisions continue for only two or three cycles, after which the larger cells drop out of the cleavage cycle.

The first asymmetric divisions in *V. carteri* occur at the transition from 32 to 64 cells, so that in an optimal situation 16 cells divide asymmetrically and 16 gonidia are produced. However, in any cultured population of *V. carteri* asexual spheroids, there is a significant percentage of spheroids with less than 16 gonidia. They result from embryos in which some anterior hemisphere cells divide symmetrically to produce only somatic cells (Jaenicke and Gilles, 1982).

There are many naturally occurring variations in the pattern of asymmetric cleavages in *V. carteri,* such as the type described above and the sexual developmental patterns (Starr, 1969). Still other variations are seen in cleavage-mutants (Kirk et al., 1991). However, in all cases the distribution of reproductive cells is directly correlated to the pattern of asymmetric divisions. Consider these illustrative examples: in sexually induced male strains, asymmetric cleavage is delayed until the final cleavage (8th or 9th division cycle), at which time all cells divide asymmetrically to yield a 1:1 ratio of somatic cells to androgonidia, which ultimately divide to produce sperm packets (Starr, 1969). In the *mul B* (multiple gonidia) mutant, as many as 32 gonidia are produced because the first asymmetric cleavages occur at the seventh rather than the sixth cleavage cycle, meaning there are 32 cells in the anterior hemisphere that can potentially divide asymmetrically (Huskey et al., 1979). Another very interesting mutant is *gls* (gonidialess), where no asymmetric cleavages occur and no gonidia are produced (Tam and Kirk, 1991).[1]

The various distributions of reproductive cells in *V. carteri* spheroids provide evidence that reproductive cell potential is not localized to any

[1]*Gls* strains which abolish asymmetric cleavages entirely can only be recovered on a genetic background containing a second mutation known as somatic regenerator (*regA*), which allows somatic cells to redifferentiate as reproductive cells (Huskey and Griffin, 1979). For a complete review of *V. carteri* cleavage-variants see Kirk et al., 1991.

particular cytoplasmic region in 32-cell embryos. Apparently, any cell above a certain threshold cell volume at the completion of embryogenesis will differentiate as a reproductive cell. Therefore, in *V. carteri* the asymmetric divisions appear to play a causal role in the segregation of reproductive cell potential. Interestingly, since activation of the pathway for reproductive cell differentiation in *V. carteri* requires a relatively large cell volume, two mechanisms function to ensure the production of large cells: asymmetric cleavages, followed by early cessation of the division cycle. This is clearly a different strategy than that seen in *V. obversus,* where reproductive cells can only be formed from the anteriormost cytoplasmic regions. However, both species show a dependence on cell volume for expression of the program of gonidial differentiation.

SPECIFICATION IN OTHER MERRILLOSPHAERA

No experimental investigations of reproductive cell specification have been carried out on other Merrillosphaera species. Nevertheless, from the descriptive accounts it is known that, even in those species where all cells are the same size at the end of embryogenesis, cells begin to differentiate as gonidia within hours. Thus, the differentiation of reproductive cells at an early stage in the life cycle is a common trait of all Merrillosphaera species. Given what has been learned about the role of cell volume in *V. carteri* and *V. obversus,* it is reasonable to speculate that *V. gigas* and *V. tertius,* which are reported to have some larger cells at the completion of cleavage, may make use of this difference in volume in specifying cells as prospective gonidia. However, size differences are completely lacking at the end of cleavage in *V. powersii* and *V. spermatosphaera* (Fig. 1C) (Vande Berg and Starr, 1971). These species must specify their prospective gonidia by a mechanism that is independent of cell volume, either by using some global patterning mechanism that involves cell–cell interactions, or by using "gonidial determinants" that are localized and segregated to specific cells. Which of these two possible mechanisms actually functions in embryos of these species could be distinguished by applying the blastomere isolation technique described here.

REFERENCES

Goldstein M (1964): Speciation and mating behavior in *Eudorina*. J Protozool 11:317–344.

Huskey RJ, Griffin BE (1979): Genetic control of somatic cell differentiation in *Volvox*. Analysis of somatic regenerator mutants. Dev Biol 72:226–235.

Huskey RJ, Griffin BE, Cecil PO, Callahan AM (1979): A preliminary genetic investigation of *Volvox carteri*. Genetics 91:229–244.

Jaenicke L, Gilles R (1982): Differentiation and Embryogenesis in *Volvox carteri*. In Jaenicke L (ed) "Biochemistry of Differentiation and Morphogenesis." pp 288–294. Springer-Verlag, New York.

Karn RC, Starr RC, Hudock GA (1974): Sexual and asexual differentiation in *Volvox obversus* (Shaw) Printz, Strains WD3 and WD7. Arch. Protistenk. 116: 142–148.

Kikuchi K (1978): Cellular differentiation in *Pleodorina californica*. Cytologia 43: 153–160.

Kirk DL, Kaufmann MR, Keeling RM, Stamer KA (1991): Genetic and cytological control of the asymmetric divisions that pattern the *Volvox* embryo. Dev Suppl 1: 67–82.

Kockert G (1968): Differentiation of reproductive cells in *Volvox carteri*. J. Protozool. 15: 438–452.

Nozaki H (1988): Morphology, sexual reproduction and taxonomy of *Volvox carteri* f. *kawasakiensis* f. nov. (Chlorophyta) from Japan. Phycologia 27: 209–220.

Pall M (1975): Mutants of *Volvox* showing premature cessation of division: evidence for a relationship between cell size and reproductive cell differentiation. In McMahon, Fox (eds): "Developmental Biology: Pattern Formation and Genetic Regulation. ICN/UCLA Symposium on Molecular and Cellular Biology, Vol. 2: 148–156." W.A Benjamin Inc., Menlo Park, CA.

Pocock MA (1938): *Volvox tertius* Meyer, with notes on two other British species of *Volvox*. J. Quekett Micros. Club, series 4, 1: 33–58.

Powers JH (1908): Further studies in *Volvox* with descriptions of three new species. Trans Amer Microsc Soc 28: 141–175.

Ransick A (1988): Experimental Analysis of Germ Cell Lineage Determination in *Volvox obversus* Embryos. Ph.D. Thesis, Univ. of Texas at Austin, pp.160.

Ransick A (1991): Reproductive Cell Specification during *Volvox obversus* Development. Dev Biol 143: 185–198.

Smith GM (1944): A comparative study of the species of *Volvox*. Trans Am Microsc Soc 63: 265–310.

Starr RC (1969): Structure, reproduction and differentiation in *Volvox carteri,* f. *nagariensis* Iyengar, strains HK 9 & 10. Arch Protistenk 111: 204–222

Starr RC (1970): Control of differentiation in *Volvox*. Dev Biol Suppl 4: 59–100.

Tam L-W, Kirk DL (1991): The program for cellular differentiation in Volvo carteri as revealed by molecular analysis of development in a gonidialess / somatic regenerator mutant. Dev 112: 571-580.

Vande Berg WJ, Starr RC (1971): Structure, reproduction and differentiation in *Volvox gigas* and *Volvox powersii*. Arch Protistenk 113: 195–219.

Evolutionary Conservation of Developmental Mechanisms, pages 71–84

6. The Evolution of Genes Regulating Developmental Commitments in Insects and Other Animals

Richard W. Beeman, Susan J. Brown, Jeffrey J. Stuart[1], and Rob Denell

USDA, ARS, U.S. Grain Marketing Research Laboratory, Manhattan, Kansas 66502 (R.W.B., J.J.S.); Division of Biology, Kansas State University, Manhattan, Kansas 66506 (S.J.B., R.D.)

INTRODUCTION

Evolutionary biologists have presented two sharply contrasting arguments for the mechanisms by which higher taxa such as orders, classes, and phyla have arisen. The strict neo-Darwinian view (e.g., Mayr, 1963) maintains that the origins of higher taxa can be explained by prolonged quantitative, polygenic changes accompanied by the extinction of intermediate forms. Alternatively, workers such as Goldschmidt (1940) emphasized the likely importance of saltational changes. Although Goldschmidt's proposals for the specific mechanisms of individual and populational genetic alterations can be firmly rejected, recent results show the importance of specific regulatory genes to the establishment during early embryogenesis of the basic body plan and have heightened support for the idea that the rare fixation of mutations of these genes could have resulted in major morphological alterations important to large-scale evolution (e.g., Raff and Kaufman, 1983; Arthur, 1988).

We know more about the genetic control of early development in the fruit fly, *Drosophila melanogaster* than any other higher animal. In this organism embryogenesis results in a larva organized along the anterior-posterior axis into a linear array of specialized metameres. This organization depends on a number of maternally and zygotically expressed genes which control the establishment of segmentation, and on eight homeotic selector genes which confer developmental commitments on each segment thus created (Akam, 1987; Ingham, 1988). Mutations in these homeotic genes yield phenotypes in which particular embryonic regions follow inappropriate developmental pathways. The homeotic selector genes are organized into two small clusters: the Antennapedia complex or ANT-C (Kaufman et al., 1990) and the bithorax complex or BX-C (Duncan, 1987). (Note, however,

[1]Current address: Department of Entomology, Purdue University, West Lafayette, Indiana 47907.

that, on the basis of studies of the red flour beetle, Beeman (1987) argued that the ancestral organization is a single homeotic complex with the ANT-C and BX-C juxtaposed, a view now widely accepted.) Their encoded proteins play a regulatory role by acting as transcription factors to control directly or indirectly the developmentally downstream genes responsible for cell differentiation and morphogenesis. This function depends on a DNA-binding region called the homoedomain, which is encoded by a highly conserved motif termed the homeobox (Scott et al., 1989). Recent work has shown that homeotic complexes are very ancient, and their organization has been highly conserved; recognizable complexes have been found in organisms as divergent as nematodes (Kenyon and Wang, 1991) and mammals (Akam, 1989).

In the light of our current knowledge, we can now refine Goldschmidt's suggestions in the form of a hypothesis that mutations in genes playing key regulatory roles in early embryogenesis have been important to the origins of higher animal taxa. The homeotic genes are particularly attractive candidates. They are known to play important roles in regulating the embryonic organization of Drosophila and vertebrates, and the high degree of sequence conservation of the homeobox region makes them easily accessible for comparative molecular studies in diverse animals. On the other hand, the high variability of the nonhomeobox portion of these genes, and their presence in animals of widely varying body plans, suggest that their function has been considerably modified in various phylogenies.

We have chosen to compare the genes in the homeotic complexes of Drosophila with those of the red flour beetle, *Tribolium castaneum*. The strongest rationale for choosing Tribolium is that it shares with Drosophila the possibility of sophisticated developmental genetic and molecular analyses, and thus will allow us to directly assess functional as well as structural aspects, and even to recognize developmentally significant genes by criteria other than homology to those important to Drosophila. In addition, these two holometabolous insects show what we anticipate is a useful degree of evolutionary divergence. That is, they are sufficiently closely related that we believe that direct comparisons will be meaningful, while at the same time they show important morphological specializations which may correlate with differences in homeotic gene function. We will address below the question of the extent to which the function of the genes of the Drosophila ANT-C reflect these specializations.

THE HOMEOTIC COMPLEXES OF DROSOPHILA AND TRIBOLIUM

Figure 1 shows the genes known in the ANT-C [Kaufman et al., 1990] and the BX-C (Duncan, 1987) of Drosophila, aligned with known or likely homologous genes thus far identified in Tribolium. Genetic studies have

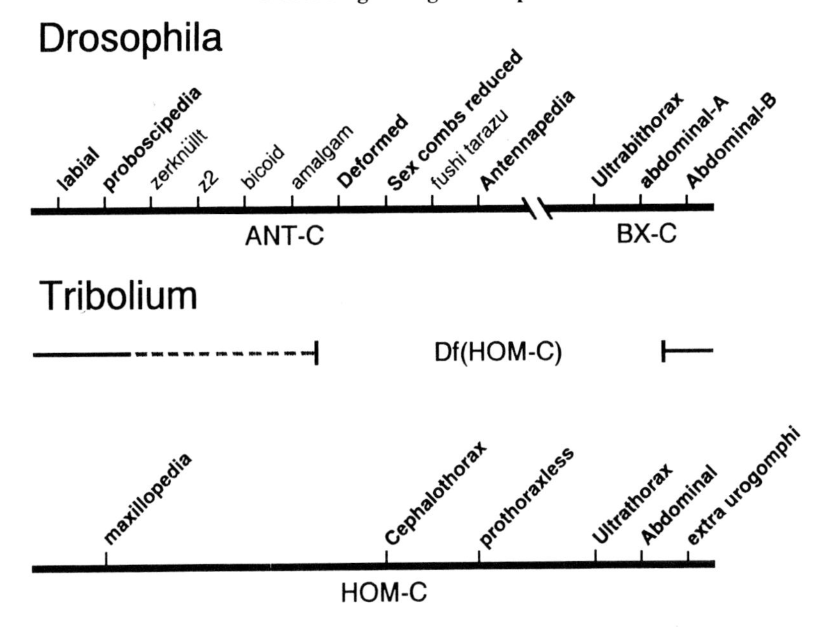

Fig. 1. A map of the HOM-C of Triboloim aligned to maps of the Drosophila ANT-C and BX-C. Homeotic genes are printed in bold, and presumed homologs are printed on the same vertical. The extent of a deficiency of the HOM-C is also indicated, with a dashed line indicating uncertainties as to its extent with respect to possible homologs of Drosophila genes in the *zerknüllt-amalgam* interval (from Stuart et al., 1991a).

identified three classes of lethal mutations within the BX-C, which correspond to three transcription units: *Ultrabithorax, Abdominal-A* and *Abdominal-B*. These homeotic genes appear to be the only protein-coding transcription units within the complex, but they are associated with cis-regulatory regions that are very large and complex. As described in more detail below, these genes regulate determinative events in the posterior thorax, abdomen, and tail regions. The ANT-C is genetically much more diverse. In addition to five homeotic genes (*labial, proboscipedia, Deformed, Sex combs reduced, and Antennapedia*) it includes those involved in other developmental processes, such as the maternal coordinate gene *bicoid* and the segmentation gene *fushi tarazu*.

The Tribolium genes were defined by spontaneous mutations (Beeman 1987), or by variants isolated in F_1 screens by dominant phenotypes or failure to complement existing lesions (Beeman et al., 1989). We know from the history of Drosophila homeotic mutations that this approach gives a biased and incomplete assessment of the functions present, and we are now engaged in saturation mutagenesis of this region. We have strong direct evidence for the homology of *Antennapedia* and *prothoraxless* (R.W.B., R. Garber and R.D., manuscript in preparation) and of *abdominal-A* and *Abdominal* (Stuart el al., 1991b). The other putative homologies indicated in Figure 1 are based on mutant phenotypic similarities. Tribolium offers a facile system for molecular studies, and in addition to developmental genetic approaches we have molecularly cloned at least a portion of the homologs of all of the Drosophila homeotic genes but *Sex combs reduced and proboscipedia.* Athough the molecular characterization of these genes is in progress, we will not emphasize that aspect of our work here. Rather, we will address two particular areas: An account of the characterization of the *Abdominal* gene, which is the best described thus far, and an evaluation of the extent to which the advanced nature of embryonic development in the diptera is reflected in the content and function of the ANT-C.

ABDOMINAL IS THE TRIBOLIUM ABDOMINAL-A HOMOLOG

We have isolated and characterized both genomic and cDNA clones from the 3' portion of the Tribolium homolog of the Drosophila BX-C gene *abdominal-A (abd-A)* (Stuart et al., 1991b). Sequence analysis shows that the encoded proteins are highly conserved in the homeodomain and flanking regions, and in the C-terminal portions. Further, the observation that a mutation at the HOM-C gene *Abdominal (A)* has a DNA rearrangement breakpoint within the protein coding region of the molecularly defined *abdominal-A* homolog demonstrates that these two homeotic genes are homologous.

The lethal syndromes associated with these two genes are strikingly similar. Like other homeotic selector genes functioning within the trunk, *abdominal-A* affects morphological units called parasegments (Martinez-Arias and Lawrence, 1985), which are comprised of the posterior compartment of one segment and the contiguous anterior compartment of the next. In Drosophila, *abd-A* mutant embryos show strong transformations of parasegments 7–9 to PS6 (posterior third thoracic segment = T3/anterior first abdominal segment = A1), as well as some subtle effects posteriorly through PS13 (Sánchez-Herrero et al., 1985; Tiong et al., 1985; Karch et al., 1985; Busturia et al., 1989). Transcripts of *abd-A* are abundant in PS7–12 and detectable in PS13–14, while the corresponding protein accumulates within the same domain (Harding et al., 1985; Regulski et al., 1985; Rowe and Akam, 1988). Tribolium embryos homozygous for *A* mutations resemble *abd-A* mutant in displaying abdominal transformations to PS6; these beetle embryos show reiteration of a parasegment which elaborates the posterior portion of the T3 larval leg and the anterior portion of the A1 pleuropodium (a glandular organ) (Stuart et al., 1991b). In situ hybridization confirms that the anterior limit of *A* expression is PS7, as it is for the expression of homologous genes in the grasshopper (Tear et al., 1990) and in the moth *Manduca sexta* (Nagy et al., 1991). This phenotype is noteworthy in two ways. It is the first demonstration of the functional significance of parasegmental expression in non-Drosophilid insects. Moreover, although these Drosophila and Tribolium larvae show reiteration of the same parasegmental domain, beetle larvae develop structures (part of the thoracic leg and pleuropodium) that do not exist in Drosophila larvae. Thus, these homologous homeotic selector genes can act to regulate remarkably diverse downstream developmental events.

The homeotic phenotypes of *abdominal-A* and *Abdominal* mutants show another remarkable difference: whereas in the former only the anterior three abdominal parasegments are well transformed, A mutants display strong transformations through the eighth abdominal segment (Stuart et al., 1991b). Interestingly, Akam et al., (1988) speculated that among primitive insects an *abdominal-A* homolog would be important to decisions in the abdomen (A2–A7 or PS7–PS13), whereas an *Abdominal-B* homolog would be important to a more posterior tail region. Our observation of a strong *Abdominal* homeotic phenotype through A8 (PS13) is largely consistent with their view of the likely primordial *abdominal-A* function in the insect lineage. We speculate that regulatory regions which control *abdominal-A* function in the posterior abdomen in *Tribolium* evolved to control *Abdominal-B* expression in the Drosophila lineage (see Stuart et al., 1991b).

IS THE ORGANIZATION AND FUNCTION OF THE ANT-C DERIVED?

Drosophila and higher diptera are highly advanced with respect to a number of aspects of embryonic development. In general, insect embryos share a conserved embroyonic stage called the segmented germband, in which the elongated embryo shows segmentation of both ectoderm and mesoderm (Sander, 1983). However, the events leading to this stage vary dramatically. In short-germ insects (such as the grasshopper), the germband initially consists of the head primordia and a subcaudal growth zone which subsequently gives rise to the metameric portion of the embryo. Embryogenesis in short-germ insects resembles that in annelids and myriapods, and is thought to be primitive. Drosophila typifies the other extreme: long-germ insects in which the primordia for all germband segments are present initially in the approximately the same proportion as after segmentation. Germband formation in some other insects, such as Tribolium, is said to be intermediate.

Drosophila also shows advanced features of anterior development. During embryogenesis there is a major morphogenetic reorganization in which most of the pregnathal and gnathal segments involute through the presumptive mouth. Thus, except for sense organs and other structures immediately surrounding the mouth, the larva is essentially headless (Turner and Mahowald, 1979). The adult head is formed from imaginal disks set aside during early embryogenesis. It, too, has undergone considerable morphological modification, which is reflected in the loss and/or fusion of imaginal primordia from different head segments.

There is no evidence that homeotic complexes from species other than diptera include homologs of the nonhomeotic genes found in the ANT-C. Thus, it is of interest to ask when these functions arose during insect evolution, and whether some may directly reflect advanced features of fly development. We anticipate that molecular and genetic studies will generate a description of the repertoire of the nonhomeotic gene functions (if any) in the HOM-C of Tribolium. At the moment, we have only one indirect piece of evidence. We have identified a deficiency of a major portion of the HOM-C; its extent, based on a combination of genetic and molecular evidence, is indicated in Figure 1 (Stuart et al., 1991a). This deficiency would be expected to remove a homolog of the segmentation gene *fushi tarazu (ftz)* if one were present at the same position in the Drosophila and Tribolium complexes. However, lethal embryos homozygous for this deficiency do not display a developmental loss of alternating regions (pair-rule phenotype) typical of *ftz* mutants in Drosophila. This observation indicates that no beetle gene at this cytogenetic location is necessary for normal segmentation. We have thus far also failed to isolate a *ftz* homolog from elsewhere in the beetle genome, and no such homolog has yet been recognized in molecular studies of honey bees (Walldorf et al., 1989) or other nondipterans. In Drosophila normal *fushi tarazu* gene function is important to development of the central nervous system as well as to segmentation, and may represent a neural gene which was coopted for a role in establishing metamerism (Doe et al., 1988). Thus it will be quite interesting to determine whether or not Tribolium has a *ftz* homolog in the HOM-C (or elsewhere) with a neural function.

The ANT-C includes five homeotic selector genes which regulate differentiative decisions in the head and anterior thorax (Kaufman et al., 1990). Several details about them suggest that their functions may have been altered during the evolution of advanced anterior development in the diptera. First, three of these genes, *labial (lab)*, *proboscipedia (pb)*, and *Deformed (Dfd)*, have no larval homeotic phenotype. Although the *pb* gene is expressed in the maxillary and labial lobes of the epidermis during embryogenesis, it does not appear to play any role important to normal development. The *lab* and *Dfd* genes are expressed in the presumed intercalary and in the mandibular/maxillary regions of the epidermis, respectively. Each of these genes does perform a vital function during embryogenesis, but their lethal syndromes do not include overt homeotic transformations of head to thoracic structures in mutant adults or homozygous adult clones.

The *Tribolium maxillopedia (mxp)* gene was first recognized by a spontaneous mutant allele which in the homozygous condition leads to the homeotic conversion of adult maxillary and labial appendages to legs (Hoy, 1966). The analysis of additional mutant alleles has shown that this original variant is hypomorphic, and in the homozygous condition putative null alleles lead to the death of larvae displaying a similar homeotic transformation as shown in Figure 2 (Beeman et al., 1989 and unpublished results). The observations that *proboscipedia* resembles *mxp* in causing a transformation of labial palps to legs (Kaufman, 1978) and that it maps to a similar position in relation to other homeotic genes argue that these genes are homologous. The embyronic function of *maxillopedia* thus suggests strongly that the ancestral *proboscipedia* gene functioned during embryogenesis, and that this role has been lost in the lineage leading to Drosophila.

We have no strong candidates for Tribolium mutations in a *Deformed* homolog, but we have cloned the homeobox-containing portion of the beetle gene and used RFLP mapping to localize it between *maxillopedia* and *prothoraxless* (Stuart et al., 1991a). As indicated in Figure 1, we have isolated a deficiency which deletes the *Deformed* homolog but not *maxillopedia* (or presumably the *labial* homolog); we don't know as yet if the nonhomeotic genes in this region of Drosophila exist in beetles or are affected by this lesion. Embryos homozygous for this deficiency chromosome are lethal, and show a transformation of all gnathal, thoracic, and abdominal segments 1–8 to antenna (Fig. 3; see Stuart et al., 1991a). The transformation of the maxillary and mandibular segments suggests that the ancestral *Deformed* gene function is necessary for correct determinative events in these segments, and that the role of the Drosophila homolog is derived.

The *prothoraxless (ptl)* locus in Tribolium was originally recognized as a spontaneous, incompletely recessive mutation causing a variable reduction of the adult prothorax (Lasley and Sokoloff, 1960). As will be reported in detail elsewhere (Beeman, Garber, and Denell, manuscript in preparation), we have isolated and characterized additional variants associated with more extreme phenotypes, and have shown that the original variant is hypomorphic. The reduction in adult T1 is a haplo-insufficient effect which can be complemented by an extra wild-type dose of the gene. Individuals homozygous for strong alleles die around the time of larval hatching, and show a strong reduction in the size of the thoracic region; this is, accompanied by a reiteration of labial pattern elements dorsally and a transformation of legs to resemble antennae ventrally. Interestingly, evidence suggests that *prothoraxless* is homologous to the Drosophila gene *Antennapedia*. Although *Antennapedia* was originally named for gain-of-function variants transforming the adult antenna to leg, the loss-of-function phenotype is an embryonic transformation of PS4 and PS5 to resemble PS3 (posterior labium/anterior T1) (Martinez-Arias, 1986), an effect which at least parallels the reiteration of the labium in *ptl* homozygotes. Moreover, in adults, *Antp*$^{-}$ clones in the ventral midthorax show a transformation to antenna (Struhl, 1981; Abbott and Kaufman, 1986). Our interpretation is that the thorax-to-antennae transformation is likely to reflect the primitive role of this gene, and that it has been retained during fly imaginal

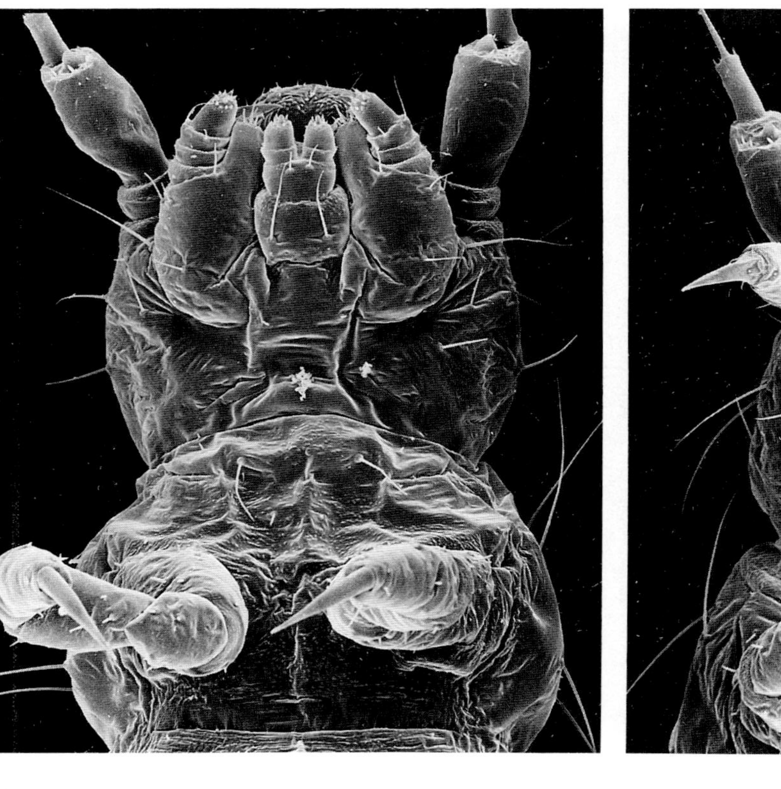

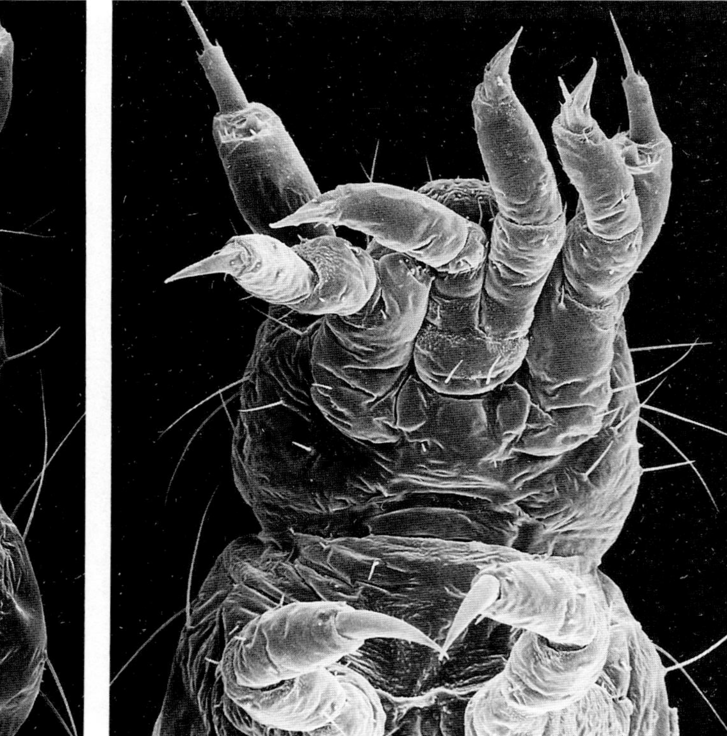

Fig. 2. A comparison of a wild-type (left) and a maxillopedia (right) larva. Scanning electron photomicrographs depict the head and prothorax of each larva in ventral view; in the mutant larva the maxillary and labial appendages of the head are transformed to thoracic legs.

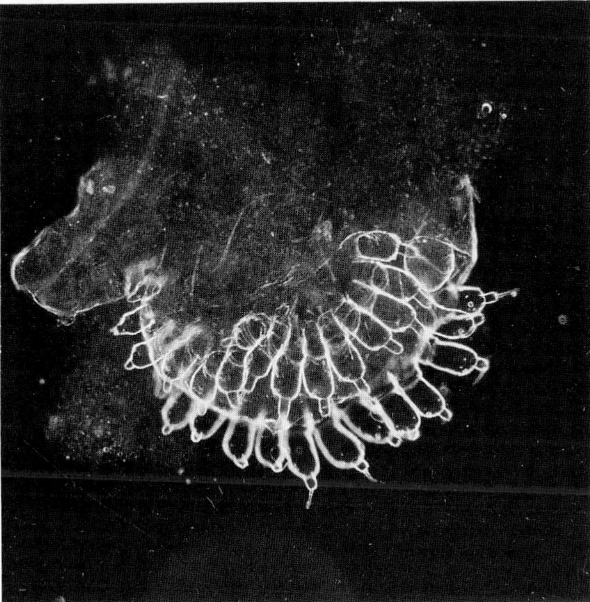

Fig. 3. A comparison of a newly hatched wild-type first instar larva (top) and a lethal embryo (unhatched larva) homozygous for Df(HOM-C). In the mutant larva, all gnathal, thoracic, and abdominal segments develop antennal appendages. Each specimen is photographed under dark-field illumination and is oriented with dorsal up and anterior to the right; the mutant larva is shown at a higher magnification (from Stuart et al., 1991a).

development but altered during embryonic development. Again, if true, this interpretation indicates that during the evolution of the diptera the downstream developmental events regulated by the *Antennapedia* gene have been greatly modified.

Based on our knowledge of the HOM-C genes deleted by the deficiency diagrammed in Figure 1, we can easily rationalize the phenotype of deficiency homozygotes. That is, loss of *Abdominal* function causes a transformation of abdominal parasegments to PS6 as described, and *Ultrathorax* mutations change A1 to thorax (unpublished observations). Transformation of the gnathal appendages to thorax can be explained by the loss of *Cephalothorax* (the putative *Sex combs reduced* homolog) and *Deformed*. Finally, as just described, mutants of *prothoraxless* cause thoracic segments to be transformed to antennae. Thus, if the gnathal and abdominal segments were transformed to thorax, and the thorax to antennae, we would see a reiteration of segments with an antennal identity as observed. We have speculated that this deficiency phenotype might be implying something important about the evolution of homeotic complexes (Stuart et al., 1991a). We suggest that, as an ancient event preceding the origin of such complexes, an *Antennapedia*-like function arose which acted positively in the trunk to distinguish it from head. By duplication and divergence of function, this primordial *Antennapedia*-like gene could have generated others in the homeotic complex specifying region-specific development within the trunk.

OVERVIEW

Our current view, based largely on comparative studies of insects and vertebrates, is that a single ancestral homeotic complex arose before the separation of the protostomes and deuterostomes, possibly by tandem duplication of a primordial gene (Kappen et al., 1989). With the notable exception of Drosophila the integrity of the ancestral complex appears to have been well conserved, although complexes appear to undergo internal expansion by further duplication and divergence, as well as by undergoing duplication of the entire complex, as indicated by the four clusters typical of vertebrates (Schughart et al., 1989).

What can account for the unprecedented evolutionary stability of these complexes? The function of the encoded proteins as transcription factors must be an important aspect. Nevertheless, genes with related homeoboxes and presumed regulatory roles have clearly been evolutionarily successful in genomic solitude. The key to the maintenance of the integrity of these homeotic complexes must surely lie in the phenomenon Lewis (1978) termed colinearity. He noted that along the anterior-posterior axis, the domains within which the genes of the BX-C of Drosophila are most important parallel the chromosomal order of the genes themselves. The same rule largely applies to the genes of the ANT-C (Kaufman et al., 1990) and all available evidence says it is valid for Tribolium, which has a single homeotic complex (Beeman, 1987). Moreover, the rule is also fulfilled with respect to the expression of each of the mouse clusters in the central nervous system and somatic mesoderm (see Reid, 1990). Most intriguing, however, is the observation that the genes of the mouse Hox-4 cluster are also expressed in a colinear fashion during limb morphogenesis (Dollé and Duboule. 1989). This graded pattern of limb expression doesn't correspond precisely to the proximal-distal or anterior-posterior axes, and shows a strong temporal as well as spatial progression. We can speculate that the most important aspect of homeotic complexes is that they include genes encoding transcription factors which can be expressed in a linear fashion spatially and/or temporally, but that the developmental significance of this patterned regulation has been modified broadly during animal evolution and can even be utilized in more than one context within a species. It was already clear from comparisons of insects and vertebrates that the genes in these complexes could regulate very diverse downstream events, and our results with the Tribolium *abdominal-A* homolog show that this conclusion can apply even to different insect orders. Moreover, the genes of insect and vertebrate homeotic complexes are clearly being regulated by very different positional cues, and their cis-regulatory elements are probably highly divergent. Thus what appears constant is the pattern of regulation, and it must be that this pattern depends on the genes being contiguous. One possibility is that the intergenic regions contain regulatory elements which affect more than one gene, so that any rearrangement with one breakpoint outside of the complex disrupts the regulation of at least one gene. Some direct evidence for this idea exists in Drosophila, where the cis-regulatory region *infraabdominal-5* has an influence on both *abdominal-A* and *Abdominal-B* function (Celniker et al., 1990). Further, in the mouse some or all Hox genes have cis-regulatory elements which can act autonomously (i.e., outside the context of the complex) to drive expression of the genes in a subset of their normal pattern. However, in no case has the entire pattern of expression been reproduced (Schughart et al., 1991). (Note that even if such an example is found, it does not necessarily disprove the hypothesis unless the corresponding region can be eliminated from the complex without disrupting the function of the remaining genes.) A second possible explanation for maintaining the integrity of the complexes, not necessarily exclusive of the one just discussed, is that a regulatory mechanism acting via spatial differences in chromatin states requires the regulated genes to be linearly arrayed. For instance, in their "open for business" model, Peifer et al. (1987) suggest that along the anterior-posterior axis, progressively more distal portions of the BX-C achieve a chromatin state which allows their interaction with trans-regulatory molecules and subsequent activation. The conservation of such a mechanism through evolutionary time would require the maintenance of the linear order of regulated genes, even though the cis-regulatory elements and functions of the genes themselves could be labile.

Given these arguments, how can we explain the divided complex of Drosophila? One speculation, rather difficult to evaluate retrospectively, is that there was a rearrangement which split the complex with the duplication of regulatory elements and mechanisms necessary for the proper expression of genes in each part. Such an explanation has been presented (Scott et al., 1989) for rearrangements which split the BX-C without disrupting its function (Struhl, 1984; Tiong et al, 1985).

As the characterization of homeotic complexes from diverse animals continues at a rapid pace, we should soon have a much clearer idea as to the genetic inventory of relatively ancient clusters and its modification in various derived lineages. It seems likely that changes in the structure and function of these complexes reflect the evolution of various animal body plans, and the hypothesis that these changes were causative factors in the origin of major taxa deserves careful evaluation.

ACKNOWLEDGEMENTS

We thank Katherine Hummels and Susan Haas for technical assistance. This work was supported by the American Cancer Center, National Aeronautics and Space Administration, National Science Foundation, US Department of Agriculture, and Wesley Foundation.

REFERENCES

Abbott MK, Kaufman TC (1986): The relationship between the functional complexity and the molecular organization of the *Antennapedia* locus of *Drosophila melanogaster*. Genetics 114:919–942.

Akam M (1987): The molecular basis for metameric pattern in the *Drosophila* embryo. Development 101:1–22.

Akam M, Dawson I, Tear G (1988): Homeotic genes and the control of segment diversity. Development 104 (Suppl.): 123–133.

Akam M (1989): *Hox* and HOM: Homologous gene clusters in insects and vertebrates. Cell 57:347–349.

Arthur W (1988): "A theory of the evolution of development." Chichester: John Wiley & Sons.

Beeman RW (1987): A homoeotic gene cluster in the red flour beetle. Nature 327:247–249.

Beeman RW, Stuart JJ, Haas MS, Denell RE (1989): Genetic analysis of the homeotic gene complex (HOM-C) in the beetle *Tribolium castaneum*. Dev Biol 133:196–209.

Busturia A, Casanova J, Sánchez-Herrero E, González R, Morata G (1989): Genetic structure of the *abd-A* gene of *Drosophila*. Development 107:575–583.

Celniker SE, Sharma S, Keelan DJ, Lewis EB (1990): The molecular genetics of the bithorax complex of *Drosophila: Cis*-Regulation in the *Abdominal-B* domain. EMBO J 9: 4277–4286.

Doe CQ, Hiromi Y, Gehring WJ, Goodman CS (1988): Expression and function of the segmentation gene *fushi tarazu* during *Drosophila* neurogenesis. Science 239:170–175.

Dollé P, Duboule D (1989): Two gene members of the murine HOX-5 complex show regional and cell-type specific expression in developing limbs and gonads. EMBO J 8:1507–1515.

Duncan I (1987): The bithorax complex. Ann Rev Genet 21:285–319.
Goldschmidt R (1940): "The material basis of evolution." New Haven: Yale University Press.
Harding C, Wedeen C, McGinnis W, Levine M (1985): Spatially regulated expression of homeotic genes in *Drosophila*. Science 229:1236–1242.
Hoy MA (1966): Section on new mutants. Tribolium Info Bull 9:85.
Ingham PW (1988): The molecular genetics of embryonic pattern formation in *Drosophila*. Nature 335:25–34.
Kappen C, Schughart K, Ruddle FH (1989): Two steps in the evolution of Antennapedia-class vertebrate homeobox genes. Proc Natl Acad Sci USA 86: 5459–5463.
Karch F, Weiffenbach B, Peifer M, Bender W, Duncan I, Celniker S, Crosby M, Lewis EB (1985): The abdominal region of the bithorax complex. Cell 43:81–96.
Kaufman TC (1978): Cytogenetic analysis of chromosome 3 in *Drosophila melanogaster*. Isolation and characterization of four new alleles of the *proboscipedia* (pb) locus. Genetics 90:579–596.
Kaufman TC, Seeger MA, Olson G (1990): Molecular and genetic organization of the Antennapedia gene complex of *Drosophila melanogaster*. Adv Genet. 27:309–362.
Kenyon C, Wang B (1991): A cluster of *Antennapedia*-class homeobox genes in a nonsegmented animal. Science 253:516–517.
Lasley EL, Sokoloff A (1960): Section on new mutants. Tribolium Info Bull 3:22.
Lewis E (1978): A gene complex controlling segmentation in Drosophila. Nature 276:141–152.
Martinez-Arias A (1986): The *Antennapedia* gene is required and expressed in parasegments 4 and 5 of the *Drosophila* embryo. EMBO J 5:135–141.
Martinez-Arias A, Lawerence PA (1985): Parasegments and comparments in the *Drosophila* embryo. Nature 313:639–642.
Mayr E (1963): "Animal species and evolution." Cambridge: Harvard University Press.
Nagy LM, Booker R, Riddiford LM (1991): Isolation and embryonic expression of an *abdominal-A-like* gene from the lepidopteran, *Manduca sexta*. Development 112:119–129.
Peifer M, Karch F, Bender W (1987): The bithorax complex: Control of segment identity. Genes Dev 1:891–898.
Raff RA, Kaufman TC (1983): "Embryos, Genes and Evolution: The Developmental Basis of Evolutionary Change." New York: Macmillan.
Regulski M, Harding K, Kostriken R, Karch F, Levine M, McGinnis W (1985): Homeobox genes of the Antennapedia and bithorax complexes of *Drosophila*. Cell 43:71–80.
Reid L (1990): From gradients to axes, from morphogenesis to differentiation. Cell 63:875–882.
Rowe A, Akam M (1988): The structure and expression of a hybrid homeotic gene. EMBO J 7:1107–1114.
Sánchez-Herrero E, Vernós I, Marco R, Morata G (1985): Genetic organization of the *Drosophila* bithorax complex. Nature 313:108–113.
Sander K (1983): The evolution of patterning mechanisms: Gleanings from insect embryogenesis and spermatogenesis. In: Goodwin BC, Holder N, Wylie CC (eds): "Development and Evolution." Cambridge: Cambridge University Press, pp 137–159.
Schughart K, Kappen C, Ruddle FH (1989): Duplication of large genomic regions during the evolution of vertebrate homeobox genes. Proc Natl Acad Sci USA 86:7067–7071.
Schughart K, Bieberich CJ, Eid R, Ruddle FH (1991): A regulatory region from the mouse *Hox-2.2* promoter directs gene expression into developing limbs. Development 112:807–811.
Scott MP, Tamkun JW, Hartzell GW III (1989): The structure and function of the homeodomain. Biochim Biophys Acta 989:25–48.
Struhl G (1981): A homeotic mutation transforming leg to antenna in *Drosophila*. Nature 292:635–638.
Struhl G (1984): Splitting the bithorax complex of *Drosophila*. Nature 308:454–457.
Stuart JJ, Brown SJ, Beeman RW, Denell RE (1991a): A deficiency of the homeotic complex of the beetle *Tribolium*. Nature 350:72–74.

Stuart JJ, Brown SJ, Beeman RW, Denell RE (1991b): The Tribolium homeotic gene *Abdominal* is homologous to *abdominal-A* of the Drosophila bithorax complex. Development (in press).
Tear G, Akam M, Martinex-Arias A (1990): Isolation of an *abdominal-A* gene from the locust *Schistocerca gregaria* and its expression during early embryogenesis. Development 110:915–925.
Tiong S, Bone LM, Whittle JRS (1985): Recessive lethal mutations within the *bithorax* complex in *Drosophila melanogaster*. Mol Gen Genet 200:335–342.
Turner FR, Mahowald AP (1979): Scanning electron microscopy of *Drosophila melanogaster* embryogenesis III. Formation of the head and caudal segment. Dev Biol 68:96–109.
Walldorf U, Fleig R, Gehring WJ (1989): Comparison of homeobox-containing genes of the honeybee and *Drosophila*. Proc Natl Acad Sci USA 86:9971–9975.

Evolutionary Conservation of Developmental Mechanisms, pages 85–110

7. Evolution of Insect Pattern Formation: A Molecular Analysis of Short Germband Segmentation

Nipam H. Patel

Carnegie Institution of Washington, Department of Embryology, Baltimore, Maryland 21210

INTRODUCTION

During the formation of all animal embryos, a basic set of coordinates is established and, concurrent with growth of the embryo by cellular proliferation, regional specifications are made and various tissues established by cellular differentiation. Experimental embryology indicates that a number of different developmental strategies are utilized to accomplish these steps.

Genetic and molecular approaches have yielded a wealth of information about the mechanisms used during early development. Early pattern formation in the embryo of the fruit fly, *Drosophila melanogaster*, has been particularly well studied. These analyses have revealed much about the processes that control early *Drosophila* development and the nature of the genes involved. Structural homologs of many of the developmentally important *Drosophila* genes have been identified in organisms from a wide variety of phyla, fueling speculation about the extent to which developmental strategies themselves have been conserved throughout evolution. Although the structural homologies among these genes argue for conserved biochemical functions, the extent to which their developmental roles are conserved is not clear.

For many of the *Drosophila* genes involved in the specification of regional identity, specifically the homeotic genes of the Antennapedia and Bithorax complexes, homologous genes have been found in other phyla. The expression patterns of these genes, such as the genes of the Hox complexes of mouse, indicate that developmental mechanisms specifying positional information along the anterior-posterior axis may be conserved among different phyla (Akam, 1989; Graham et al., 1989; McGinnis and Krumlauf, 1992). Homologs of some of the *Drosophila* genes involved in the process of segmentation have also been identified in other phyla (for examples see Kamb et al., 1989; Bastian and Gruss, 1990; Joyner and Martin, 1987; Dressler et al., 1988). For these genes, however, the extent to which they serve similar developmental functions in pattern formation in different phyla is unclear. Resolution of this question has been hindered by the difficulty of comparing embryogenesis between such highly divergent organisms as *Drosophila* and mouse.

One way to better understand the evolution of pattern formation is to study the evolution of segmentation within the arthropod lineage itself, particularly the evolution of insect segmentation. Such studies will also further our understanding of the process of *Drosophila* development since phylogenetic history is an important factor in embryonic development. In addition, by examining the evolution of insect segmentation, we will obtain a better understanding of the potential mechanisms of segmentation used by ancestral arthropods. This in turn will help to clarify our expectations of the similarities that may exist between arthropod developmental strategies and those utilized by other phyla.

Arthropod Phylogeny

The establishment of a phlyogenetic tree for arthropods has been the center of considerable debate. Manton (1977) has argued, based on morphological criteria, that arthropods should not be considered a single phylum but should be divided into three phyla: Uniramia (insects, myriopods, etc.), Chelicerata (spiders, scorpions, etc), and Crustacea (crayfish, lobsters, etc.). According to this view, these three groups independently evolved an "arthropod" body plan and represent an example of convergent evolution. More recent molecular data argues against this "polyphyletic" view (Field et al., 1988). Alternatively, Valentine (1989) suggests that the different arthropod and annelid ancestors emerged from a plexus of precursors.

Within the arthropods is the class Insecta, the largest class in the animal kingdom, with close to 1 million identified species (Schwalm, 1988). The phylogeny of the class Insecta is quite complex, but a basic outline has met with general agreement. Figure 1 presents a phylogentic tree of several of the better-known insect orders.

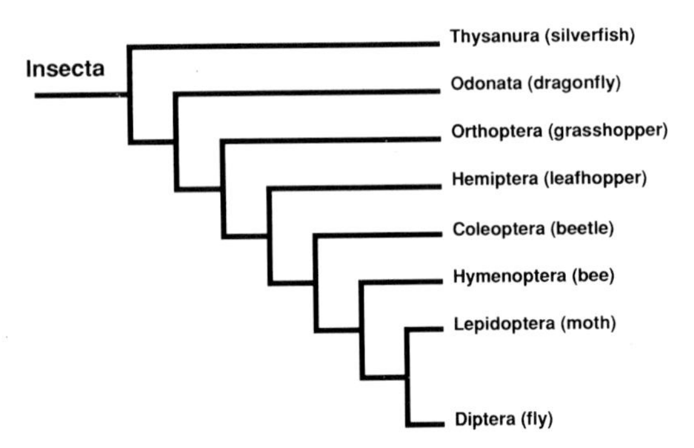

Fig. 1. Phylogenetic tree of several insect orders. (Adapted from Schwalm, 1988). Common name of a member of each order is given in parentheses.

Most of our knowledge of the molecular mechanisms of segmentation comes from studies on *Drosophila*. The phylogenetic tree, however, indicates that Dipterans are one of the most evolved of all insects in the sense that they have undergone the greatest amount of evolutionary change (Tear et al., 1988). More importantly, previous comparative embryology studies indicate that marked alterations in early development have occurred during insect evolution and that the early embryology of *Drosophila* represents a clear divergence from the mode of development seen in most other insects.

Comparative Embryology of Insect Development

All arthropod embryos, including those of insects, pass through a highly conserved, "phylotypic" point in development that is called the germband stage (Sander, 1983). This stage occurs after the embryo has completed gastrulation, when it assumes an elongated shape in which the body is overtly segmented. The way that various insect embryos reach this stage, however, varies quite dramatically. These differences can be illustrated by comparing the development of *Schistocerca* (grasshopper) and *Drosophila*, representatives of the two extremes of insect developmental strategies.

Drosophila oogenesis occurs in a meroistic type of ovary, in which nurse cells are connected by a cytoplasmic bridge to the anterior end of the egg. The nurse cells provide the developing egg with large quantities of maternal RNA. By contrast, *Schistocerca* oogenesis takes place within a panoistic ovary, in which there are no nurse cells. The RNA inside the oocyte comes from the oocyte nucleus itself (insect ovaries reviewed in King and Büning, 1985; French, 1988). These differences in ovary types may account for the differences in the relative rates of development (eggs from panoistic ovaries tend to develop much slower than those from meroisic ovaries) as well as differences in some of the earliest steps in anterior-posterior pattern formation (see below).

After the initial events of fertilization, the *Drosophila* embryo undergoes a highly synchronous series of nuclear divisions in the absence of cell divisions, thus forming a syncytium. The nuclei begin migrating toward the periphery of the egg, and shortly after reaching it, they become separated by cell membranes, resulting in the formation of a cellular blastoderm. The embryo at this stage appears to comprise a homogeneous field of cells spread across the entire egg, but a multitude of manipulations such as ligation, fragmentation, centrifugation, and localized irradiation of the blastoderm embryo indicate that the cells are anything but homogeneous (reviewed by Sander, 1976). The cells of the *Drosophila* blastoderm are already developmentally committed to contribute to specific regions of the final body plan, and thus, the *Drosophila* blastoderm embryo contains a precise representation of the complete larval animal. The same types of manipulation also indicate that positional values are specified in *Drosophila* eggs even before the blastoderm stage, leading Sander (1976) to suggest that gradients of positional information are established in the oocyte and that interactions between these gradients are able to establish the entire body plan by the blastoderm stage.

This highly determined blastoderm stage is not characteristic of the development of the grasshopper. After fertilization, the embryo develops syncytially as the nuclei divide and begin to migrate toward the periphery of the egg as in *Drosophila* (Schwalm, 1988). It should be noted that Roonwal (1936) argued that this early development was not syncytial because the migrating nuclei and their surrounding islands of cytoplasm were not interconnected by cytoplasm, but rather were separated by yolk granules. There is, however, no evidence for cell membranes around the migrating nuclei, so the description of this stage as syncytial appears valid. Cellularization occurs after the nuclei reach the periphery but differs from that in *Drosophila* in that it is only those cells at the posterior pole of the egg that give rise to the embryo itself. The remaining cells give rise to extraembryonic membranes (Moloo, 1971). The cells at the posterior end form a small disc of about 300 μm in diameter that is located within the 7-mm long egg. This disc of cells, which represents the *Schistocerca* blastoderm, differs from the highly determined *Drosophila* blastoderm in that it does not appear to contain a complete projection of the final body plan. Various observations and manipulations of the *Schistocerca* embryo which are similar to those carried out on *Drosophila* indicate that although the *Schistocerca* blastoderm embryo contains cells determined to give rise to the head, a posterior growth zone gives rise to the remaining segmented portion of the body one segment at a time during an extended phase of cellular proliferation (Mee and French, 1986; Mee, 1986). The more posterior regions are specified one segment at a time during this growth phase. Manipulations before the blastoderm stage indicate that the more posterior region of the egg contains information specifying the location for the initial blastoderm disc, but that the egg does not contain information for the specification of the different regions of the embryo (Moloo, 1971).

Embryos, such as those of *Drosophila*, that contain a complete representation of the body plan at the blastoderm stage, are classified as long germband embryos (Krause, 1939), whereas embryos, such as those of *Schistocerca*, that develop segments sequentially during a growth phase after the blastoderm stage are classified as short germband embryos (Krause, 1939; Sander, 1976). There are also insects that utilize an intermediate mode of development; such insects are said to have intermediate, or "semi-long", germband embryos. Hemipterans are an example of this intermediate germband type. The blastoderm contains information for the specification of head, gnathal, and thoracic segments, but the entire abdomen arises from a growth zone (Sander, 1976). The distribution of germband types is interesting from an evolutionary standpoint because it raises the possibility that representatives of intermediates in the evolution of short to long germband development can be found among extant insects. However, the precise nature of the relationship of intermediate-germband development to long and short germband development is unclear (see below).

The initial specification of an anterior domain with progressive addition of posterior segments is a common characteristic of the development of other arthropods, such as crustaceans, and is even comparable to the early development of annelids. Additionally, panoistic ovaries appear to be ancestral to meroistic ovaries (King and Büning, 1985). Thus, the production of eggs in a panoistic ovary and the short germband mode of embryonic development may be most similar to the developmental strategy utilized by the common ancestor of all insects and, most likely, all arthropods (Tear et al., 1988). Our understanding of the molecular and genetic basis of segmentation, however, is mostly confined to the development of the more highly evolved long-germband embryo of *Drosophila*.

Molecular Genetics of *Drosophila* Pattern Formation

Many of the genes essential for segmentation were identified in genetic screens for loci that are required for the production of the wild-type cuticular pattern (Nüsslein-Volhard and Wieschaus, 1980; Jurgens et al., 1984; Nüsslein-Volhard et al., 1984; Wieschaus et al., 1984). Subsequent analysis has revealed that the genetic hierarchy controlling *Drosophila* segmentation in the anterior-posterior axis requires maternal components and the sequential expression of zygotic gap, pair-rule, and segment polarity genes (see Akam, 1987 and Ingham, 1988 for reviews).

As stated previously, analyses of developmental commitment indicate that the cellular blastoderm is highly determined, although the embryo appears to be composed of a homogeneous field of cells. Likewise, molecular and genetic evidence indicate that the process of segmentation is well underway by the blastoderm stage. Maternal components have already established gradients of activity emanating from the two ends of the embryo before the syncytial blastoderm stage (reviewed in St Johnston and Nüsslein-Volhard, 1992). For example, *bicoid* RNA, which is supplied by the nurse cells via the cytoplasmic bridge to the anterior end of the egg, remains mostly localized at the anterior end of the egg (Frigerio et al., 1986), and the protein is found in a gradient with its highest concentration at the anterior pole (Driever and Nüsslein-Volhard, 1988). A gradient of information from the posterior end is generated by the posterior group genes, with *nanos* being the posterior determinant (Wang and Lehmann 1991). These gradients influence the expression of gap genes, which are expressed in large zones of the embryo during the syncytial blastoderm stage. Interactions among the gap genes in their areas of overlap presumably control the expression of the next class of genes, the pair-rule genes (reviewed in Small and Levine, 1991). The protein products of the pair-rule genes appear in a series of 7 stripes. Each stripe and interstripe initially has approximately the width of a final segment, and the phasing of stripes is slightly different for each gene. The complex overlapping pattern of pair-rule gene activation then determines the expression of the final class of segmentation genes, the segment polarity genes (reviewed in Ingham, 1991). The products of most of these genes appear at the cellular blastoderm stage in a pattern of 14 or 15 stripes, with a single stripe per segment. Finally, an additional class of genes, the homeotic genes, are activated in a series of overlapping regions of the embryo and serve to give specific identities to each of the segments (Lewis, 1978; Peifer et al., 1987; McGinnis and Krumlauf, 1992).

MOLECULAR COMPARISON OF LONG AND SHORT GERMBAND DEVELOPMENT

Comparative embryology indicates that there may be important differences in the mechanisms used for segmentation in long and short germband embryos. Sander (1976, 1983, and 1988) provides a general framework for the comparison of the two modes of development. For long-germband development, the embryological data can be combined with our present knowledge of the genetic and molecular basis for segmentation in *Drosophila*.

Maternal contributions to the oocyte, supplied by the nurse cells, are essential to the initial specification of the long germband embryo. These maternal influences establish the initial gradients of information, such as the *bicoid* protein gradient, within the egg. This initial information is utilized to sequentially subdivide the embryo into smaller and smaller units, which is reflected in the subsequent expression of gap and pair-rule genes. The expression patterns of the segment polarity and homeotic genes indicate that the entire body plan has been established by the end of the blastoderm stage. Thus, in long germband embryos, such as *Drosophila,* segments are formed simultaneously, rapidly, and without growth. The blastoderm is simply subdivided into smaller and smaller units by the sequential expression of the gap, pair-rule, and segment polarity genes.

Short germband embryos, by contrast, may be less dependent on maternal contributions. The egg appears to contain information only for the location of the developing blastoderm disc and for the initial polarity of growth. The generation of segments is tied to the subsequent expansion of the proliferative zone of the initial blastoderm disc. The segments appear one at a time as the embryo elongates, rather than simultaneously. Thus, metameric patterning in short germband embryos is associated with a proliferative growth zone that generates segments one at a time in a rostrocaudal gradient after the formation of the blastoderm.

The existence of these two modes of segmentation—progressive subdivision of a sheet of cells versus addition of segments one at a time—raises several questions. What mechanisms are conserved between long germband and short germband development, and what mechanisms are unique to each mode? How did simultaneous segmentation evolve from sequential segmentation? What can we infer about the evolution of the process of segmentation and its molecular and genetic basis by comparing segmentation in different arthropods?

Comparative embryology raises these questions and molecular approaches are being used to help answer them. Sequence homologies can be used to isolate *Schistocerca* homologs of *Drosophila* segmentation genes, which in turn can be used as probes to examine the process of short germband segmentation at the molecular level. *Schistocerca* homologs of the *Drosophila* homeotic gene *abdominal-A,* the segment polarity gene *engrailed,* and the pair-rule gene *even-skipped* have been isolated and their protein distribution patterns studied in detail (Patel et al., 1989a,b; Tear et al., 1990; Patel et al., 1992). These studies reveal that certain aspects of *Drosophila* pattern formation are shared by short germband insects, whereas other aspects may be unique to long germband development. Because the results with grasshopper *engrailed* were published some time ago, I will summarize those results only briefly. The analysis of *even-skipped* in grasshopper (Patel et al., 1992) is more recent and provides the clearest molecular evidence for differences between the mechanisms of pattern formation in short germband and long germband embryos.

Expression of *engrailed* in Schistocerca

Drosophila engrailed, a member of the segment polarity class of genes, is expressed in the posterior portion of each segment from the embryo to the adult, and thus can be considered as a molecular marker defining *Drosophila* segments (Kornberg et al., 1975; Fjöse et al., 1985; DiNardo et el., 1985). The expression of the grasshopper *engrailed* homolog was analyzed using the monoclonal antibody MAb 4D9, which recognizes a highly conserved epitope of *engrailed* homologs in several organisms (Patel et al., 1989b). These experiments indicated that the expression pattern of *engrailed* in long and short germband organisms is very similar at the germband stage; engrailed protein accumulates in the cells of the posterior portion of each segment. For *Drosophila,* the germband stage is from approximately stage 10 (4 h after egg laying) until stage 13 (10 h after egg laying). The grasshopper germband stage is from about 30% of embryonic development to 50% of embryonic development. Figure 2G shows the pattern of *engrailed* protein expression in a germband stage grasshopper embryo. The expression pattern of grasshopper *abdominal-A* at the germband stage also closely matches the expression pattern of *Drosophila abdominal-A* at an equivalent stage of development (Tear et al., 1990). The highly conserved patterns of *engrailed* and *abdominal-A* expression at the germband stage are consistent with the

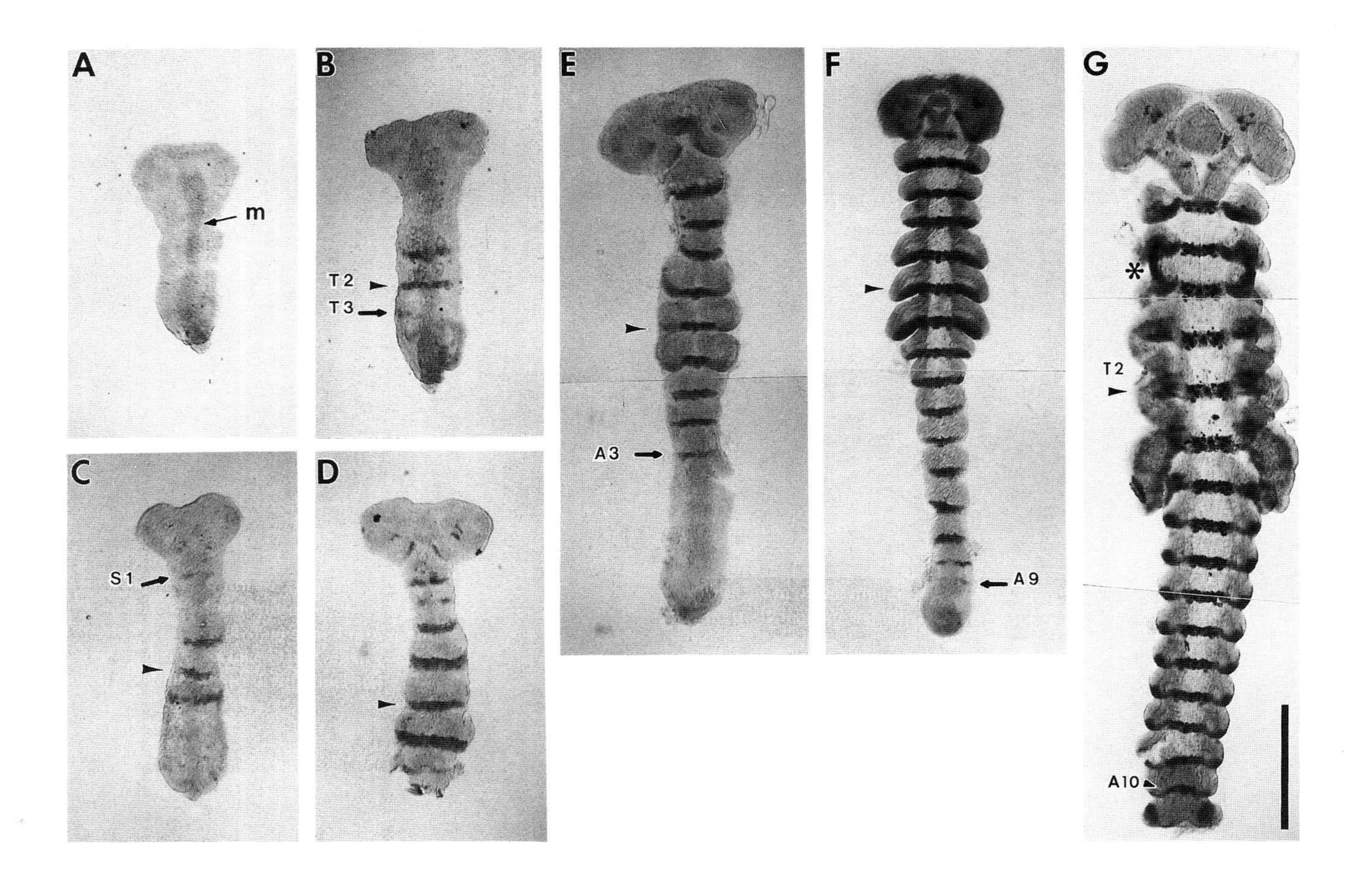

A
m
B
T2
T3
C
S1
D
E
A3
F
A9
G
*
T2
A10

hypothesis that this stage represents a highly conserved and constrained point in arthropod development.

The ways in which these expression patterns are established, however, provide molecular confirmation that the mechanisms of segmentation may differ between long and short germband insects. In *Drosophila,* the *engrailed* stripes begin to form during the blastoderm stage. They appear in a rapid anterior to posterior progression across the length of the embryo, subdividing the existing sheet of cells at regularly spaced intervals (see Fig. 6A). The stripes show a pair-rule pattern of initiation that is thought to reflect the control of *engrailed* expression by the pair-rule genes (reviewed in Ingham, 1991). This pattern of appearance of *engrailed* stripes is consistent with the embryological data indicating that the *Drosophila* blastoderm contains a complete representation of the final larval body plan. The same types of embryological studies suggest, however, that segments arise one at a time during *Schistocerca* development. This is confirmed at the molecular level by the observation that *engrailed* stripes in *Schistocerca* appear one at a time during a growth phase, rather than simultaneously at the blastoderm stage (see Figs. 2 and 3). Stripe formation begins in the region of the first and second thoracic segments during gastrulation and spreads both anteriorly and posteriorly. The sequential formation of stripes is most striking in the abdominal segments, where stripe formation follows well behind the advancement of the growth zone and about two segments ahead of the first morphologically visible signs of segmentation. At no point does the *engrailed* pattern show the existence of a compressed stripe pattern that simply expands with growth, and there is no indication of a "pair-rule" pattern to the stripe formation. Thus, the analysis of *engrailed* expression provides molecular evidence that *Schistocerca* segments are generated one at a time after the blastoderm stage. This conclusion is also supported by the initiation of *abdominal-A* protein expression. In *Drosophila, abdominal-A* protein appears almost simultaneously throughout its entire domain of expression at the germband stage. In grasshopper, however, *abdominal-A* protein first appears in an anterior abdominal region and then expands posteriorly, lagging about one segment behind the advancing *engrailed* stripes (Tear et al., 1990).

Fig. 2. Appearance of *engrailed* stripes in the grasshopper embryo. Photographs of *engrailed* staining in grasshopper embryos at various percentages of development. **A**: 15% **B**: 17% **C**: 19% **D**: 22% **E**: 23% **F**: 29% **G**: 33%. Anterior is up in all panels and the position of the second thoracic stripe (T2) is indicated by an arrowhead. All photos are at the same magnification. At 15% (A), no *engrailed* stripes are visible and gastrulation has begun to generate mesoderm (m). At 17% (B), the third thoracic stripe (T3) is just beginning to form. At 19% (C), the S1 stripe is appearing and at 19% (D) the cephalic stripes are forming. The third abdominal stripe (A3) is forming at 23% (E), and the ninth abdominal stripe (A9) appears at 29% (F). Note that the embryos in E and F are almost the same length, but the *engrailed* stripes are still progressively appearing down the length of the abdomen. By 33% (G) the final *engrailed* stripe (A10) has appeared. An asterisk indicates lateral connection between S2 and S3 stripes. Scale bar = 500 μm (from Patel et al., 1989a).

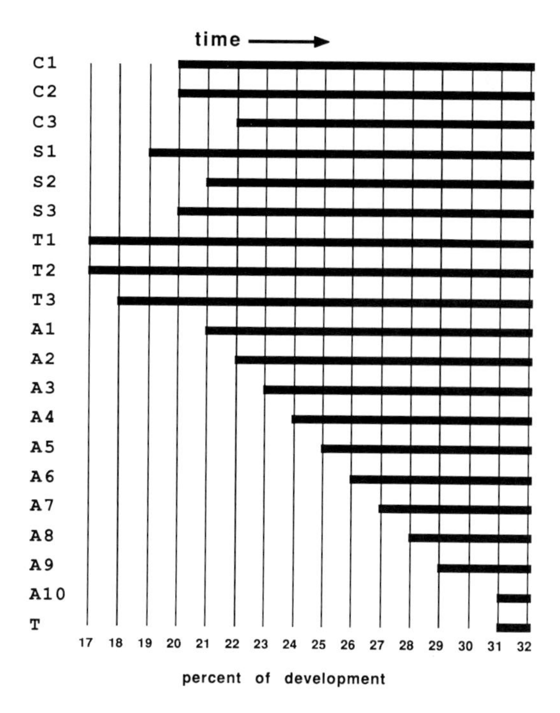

Fig. 3. Appearance and timing of the *engrailed* stripes in the grasshopper embryo. The horizontal axis represents time in terms of percent development, and the vertical axis denotes the different body segments. The spatiotemporal expression pattern matches Seidel's concept of the anterior thoracic region as the morphological differentiation center (reviewed in Sander, 1976). The abdominal *engrailed* stripes appear in a sequential pattern with one stripe appearing about every 1% of development (from Patel et al., 1989a).

Molecular evidence that the type of short germband development seen in grasshopper is ancestral to *Drosophila* long germband development comes from studies of the expression of *engrailed* during crustaceans development (Patel et al., 1989b). Crustaceans, like grasshopper, form *engrailed* protein stripes one at a time behind a proliferative zone. This observation suggests that the mode of segmentation seen in *Schistocerca,* as opposed to that in *Drosophila,* is more closely related to the process of segmentation in the common ancestor to all arthropods.

The results with *engrailed* and *abdominal-A* agree with the accumulated embryological data. The highly conserved expression patterns of these genes at the germband stage are consistent with the hypothesis that this is a highly conserved and constrained developmental point in arthropod development. The way in which these patterns are initiated, however, confirms at the molecular level that grasshopper segments are generated sequentially during a growth phase and not simultaneously at the blastoderm stage as in *Drosophila.* The differences in the initiation of segment polarity and homeotic gene expression patterns suggests that the regulation of these genes may be different in long and short germband insects. In *Drosophila,* the expression of segment polarity genes, such as *engrailed,* is under the regulation of the pair-rule genes. The analysis of *engrailed* expression in grasshopper, however, suggests that the expression of segment polarity genes might not be under the control of pair-rule genes in short germband embryos.

Expression of *even-skipped* in Schistocerca

To address this hypothesis, we cloned the grasshopper homolog of the *Drosophila* pair-rule gene *even-skipped* (*eve*) and analyzed the distribution of its protein product (Patel, et al., 1992). Previous genetic analysis showed that *even-skipped* is required to establish the proper pattern of *engrailed* stripe expression in the blastoderm embryo of *Drosophila* (MacDonald et al., 1986; Frasch et al., 1987; Frasch et al., 1988; DiNardo and O'Farrell, 1987). *eve* has a second phase of expression (Macdonald et al., 1986; Frasch et al., 1987) during the germband stage, when it is expressed in the developing central nervous system (CNS), in dorsal mesoderm, and in ectodermal cells of the anal pad. A temperature-sensitive allele of *even-skipped* has been used to show that the CNS expression plays an important role in the differentiation of particular *Drosophila* neurons (Doe et al., 1988). Because clear homologs of these neurons are readily identifiable in grasshopper, and because neurogenesis is very similar in *Drosophila* and grasshopper, we reasoned that an *even-skipped* homolog should be present in grasshopper, if only for its neural function.

A portion of the grasshopper *even-skipped* homeobox was amplified by PCR, and this fragment was used to isolate a cDNA. Figure 4 shows a comparison of the proteins encoded by the *Drosophila even-skipped* gene and the putative grasshopper homolog. A high degree of amino-acid identity exists within the homeodomain (56/60 amino acids), in the adjacent C-terminal region (region A in Fig. 4; 17/24 amino acids), and in a short stretch at the C-terminal end of the protein (region B in Fig. 4; 7/10 amino acids). No homeodomain protein from *Drosophila,* other than *even-skipped,* has more than 60% amino-acid identity with the homeodomain of the putative grasshopper *eve* homolog, and none has significant sequence similarity outside of the homeodomain. Thus, the grasshopper gene described here is most likely a homolog of the *Drosophila even-skipped* gene.

The isolated cDNA was used to produce fusion proteins, and antibodies generated against the fusion proteins allowed us to examine the pattern of *even-skipped* protein accumulation during grasshopper embryonic development. About midway through embryogenesis, *eve* is expressed in a segmentally repeated pattern of neurons, in an ectodermal ring at the anal pad, and

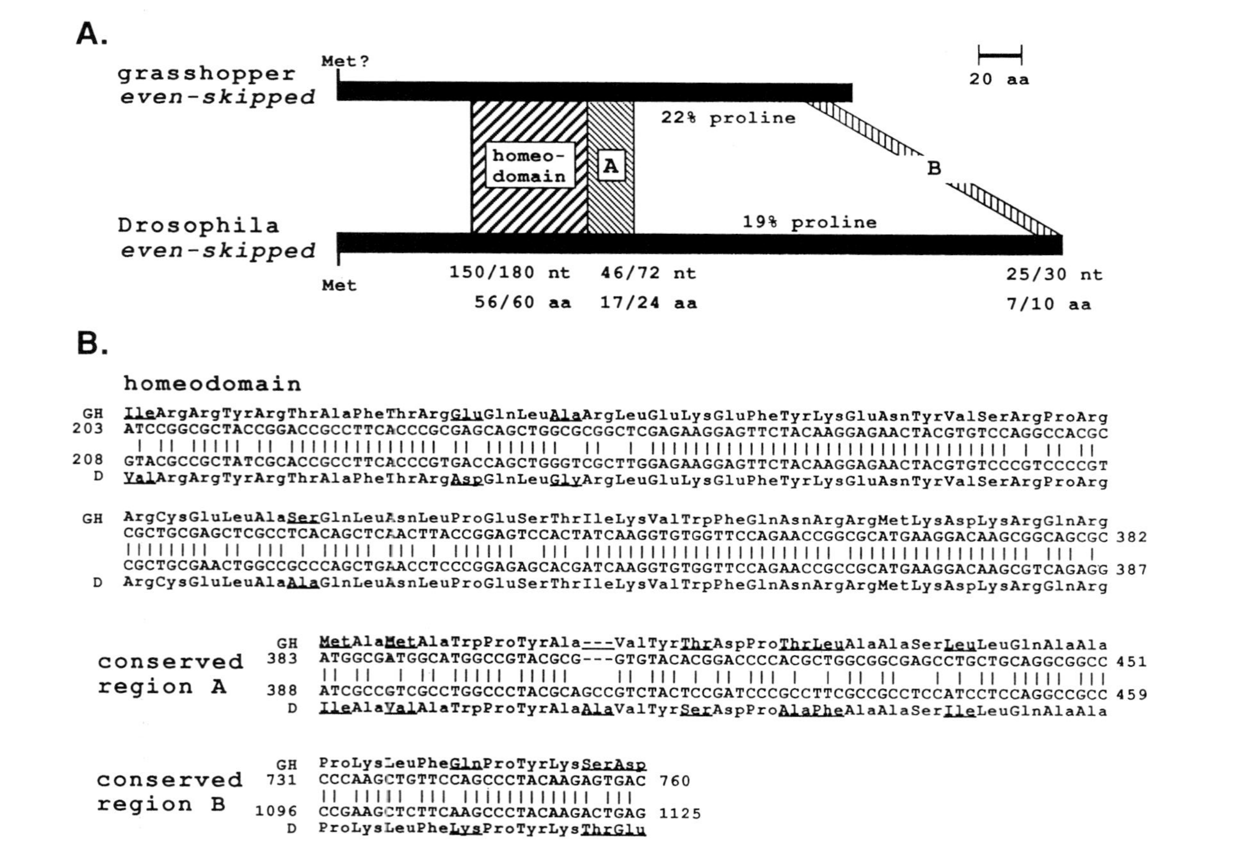
A.
grasshopper
even-skipped
Met?
homeo-
domain
A
22% proline
B
20 aa
Drosophila
even-skipped
19% proline
Met
150/180 nt
56/60 aa
46/72 nt
17/24 aa
25/30 nt
7/10 aa
B.
homeodomain
GH IleArgArgTyrArgThrAlaPheThrArgGluGlnLeuAlaArgLeuGluLysGluPheTyrLysGluAsnTyrValSerArgProArg
203 ATCCGGCGCTACCGGACCGCCTTCACCCGCGAGCAGCTGGCGCGGCTCGAGAAGGAGTTCTACAAGGAGAACTACGTGTCCAGGCCACGC
208 GTACGCCGCTATCGCACCGCCTTCACCCGTGACCAGCTGGGTCGCTTGGAGAAGGAGTTCTACAAGGAGAACTACGTGTCCCGTCCCCGT
D ValArgArgTyrArgThrAlaPheThrArgAspGlnLeuGlyArgLeuGluLysGluPheTyrLysGluAsnTyrValSerArgProArg
GH ArgCysGluLeuAlaSerGlnLeuAsnLeuProGluSerThrIleLysValTrpPheGlnAsnArgArgMetLysAspLysArgGlnArg
CGCTGCGAGCTCGCCTCACAGCTCAACTTACCGGAGTCCACTATCAAGGTGTGGTTCCAGAACCGGCGCATGAAGGACAAGCGGCAGCGC 382
CGCTGCGAACTGGCCGCCCAGCTGAACCTCCCGGAGAGCACGATCAAGGTGTGGTTCCAGAACCGCCGCATGAAGGACAAGCGTCAGAGG 387
D ArgCysGluLeuAlaAlaGlnLeuAsnLeuProGluSerThrIleLysValTrpPheGlnAsnArgArgMetLysAspLysArgGlnArg
conserved
region A
GH MetAlaMetAlaTrpProTyrAla---ValTyrThrAspProThrLeuAlaAlaSerLeuLeuGlnAlaAla
383 ATGGCGATGGCATGGCCGTACGCG---GTGTACACGGACCCCACGCTGGCGGCGAGCCTGCTGCAGGCGGCC 451
388 ATCGCCGTCGCCTGGCCCTACGCAGCCGTCTACTCCGATCCCGCCTTCGCCGCCTCCATCCTCCAGGCCGCC 459
D IleAlaValAlaTrpProTyrAlaAlaValTyrSerAspProAlaPheAlaAlaSerIleLeuGlnAlaAla
conserved
region B
GH ProLysLeuPheGlnProTyrLysSerAsp
731 CCCAAGCTGTTCCAGCCCTACAAGAGTGAC 760
1096 CCGAAGCTCTTCAAGCCCTACAAGACTGAG 1125
D ProLysLeuPheLysProTyrLysThrGlu

in a segmentally repeated pattern of dorsal mesoderm, a pattern identical to that seen in *Drosophila* at an equivalent stage of development (Fig. 5A, B). In both insects, the dorsal mesoderm cells that express *eve* go on to form part of the dorsal vessel (heart) and the most dorsal muscle fibers of the body wall.

A careful examination of the patterns of neural expression suggests that *eve* is expressed in homologous cells in the nervous systems of both insects. For example, *eve* protein is present in only the aCC, pCC, and RP2 neurons on the dorsal surface of the CNS in both insects (Fig. 5C, D). More ventrally, *eve* is expressed in both insects in the three U neurons and a pair of neurons (*CQ*) located next to the U neurons. A cluster of neurons, termed the EL neurons, also expresses *eve,* although *Drosophila* appears to have a greater number of these neurons than does grasshopper. Our detailed analysis of *eve* expression also revealed a surprising correlation between *even-skipped* expression by motorneurons and muscles. In both insects, *eve* is expressed by five motorneurons, aCC, RP2, and the three Us, and by the targets of these motorneurons, the dorsalmost muscle fibers. It is possible that *eve* regulates the expression of cell surface markers involved in the recognition of specific muscle targets by identified motorneurons. The similarities in expression pattern, particularly in the nervous system, that are found at these stages provide further evidence that the isolated grasshopper gene is the homolog of *Drosophila even-skipped.*

To analyze the potential pair-rule function of grasshopper *even-skipped,* we studied the expression patterns of *even-skipped* at much earlier stages of development. During the blastoderm stage, *Drosophila even-skipped* expression is resolved into a series of 7 stripes (Macdonald et al., 1986; Frasch et al., 1987). Shortly thereafter, *engrailed* expression is initiated in a pattern of 14 stripes, with the anterior margin of each *even-skipped* stripe coinciding with an odd-numbered *engrailed* stripe (Lawrence et al., 1987). At the onset of gastrulation, *even-skipped* stripes have sharply defined anterior borders and *engrailed* stripes show a pair-rule pattern of intensity reflecting their order of initiation (Fig. 6A, D). As mentioned previously, detailed genetic analysis has indicated that the pair-rule pattern of *even-skipped* expression is required to

---

Fig. 4. Alignment of *Drosophila*And grasshopper *even-skipped* sequences. **A**: Schematic representations of the grasshopper and *Drosophila even-skipped* proteins indicating the three regions of amino-acid sequence similarity: the homeodomain; region "A" (extended homeodomain); region "B." Below each region, the ratio of nucleotide and amino-acid identities is given. Between regions "A" and "B," the predicted sequence of both proteins indicates a high proportion of proline residues. **B**: Nucleotide and amino-acid alignments of the three regions of similarity. Amino-acid mismatches are underlined and a single amino-acid gap has been inserted in grasshopper region "A" (dashed lines) for the purposes of sequence alignment. For *Drosophila even-skipped*, nucleotide numbering (shown at the beginning and end of each domain) begins with the A of the start methionine codon of the protein. For the grasshopper sequence, numbering begins with the first nucleotide of the cDNA sequence after the EcoRI cloning site (from Patel el al., 1992).

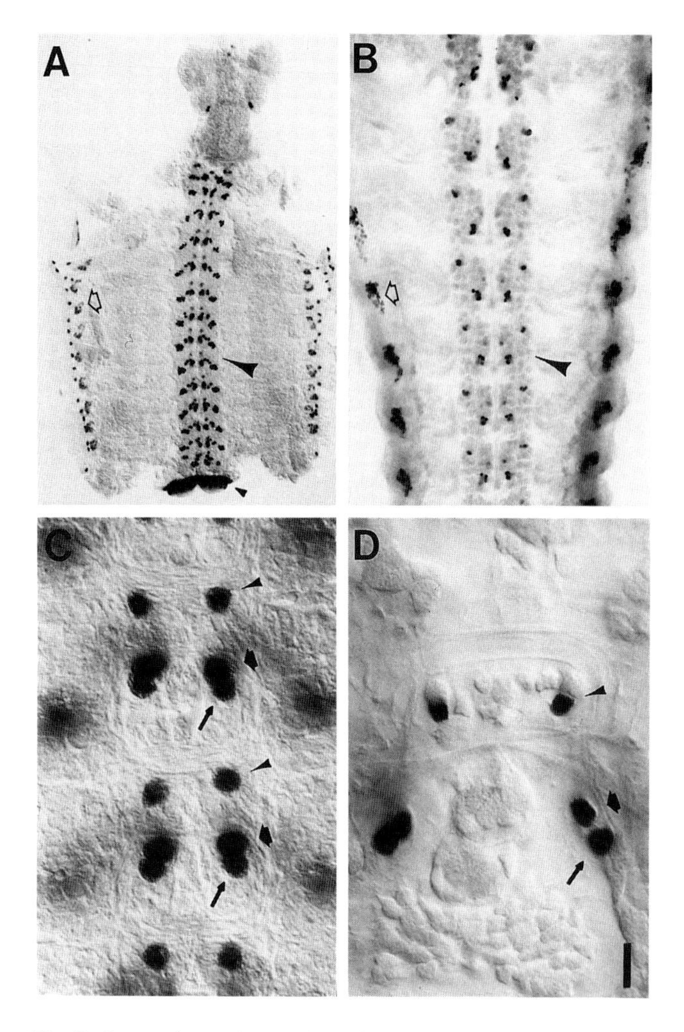

Fig. 5. Comparison of *even-skipped* expression in *Drosophila* and grasshopper at midembryogenesis. **A** and **C**: *even-skipped* expression, as visualized with MAb 3C10, in a *Drosophila* embryo at 12h of development. **B** and **D**: *even-skipped* expression in a 45% developed grasshopper embryo, as detected with an affinity-purified rat antiserum. In both *Drosophila* (A) and grasshopper (B) embryos, *even-skipped* protein accumulates in a subset of neurons (large arrowhead), in dorsal mesoderm (open arrow), and in an ectodermal ring at the anal pad (small arrowhead in A; not shown in B). The *Drosophila* embryo appears to have more neural staining, but this is simply because the greater thickness of the grasshopper embryo prevents visualizing all stained neurons in a single focal plane. Homologous neurons express *even-skipped* in both insects. For example, panels C and D show the dorsalmost neurons that express *even-skipped* in *Drosophila* (C) and grasshopper (D). In both, *even-skipped* protein is detected in RP2 (arrowhead), aCC (wide arrow), and pCC (thin arrow), but not in any of the other well-characterized dorsal neurons. Scale bar = 70 μm (A and B); 8μm (C); 10 μm (D). Anterior is up in all panels (from Patel et al., 1992).

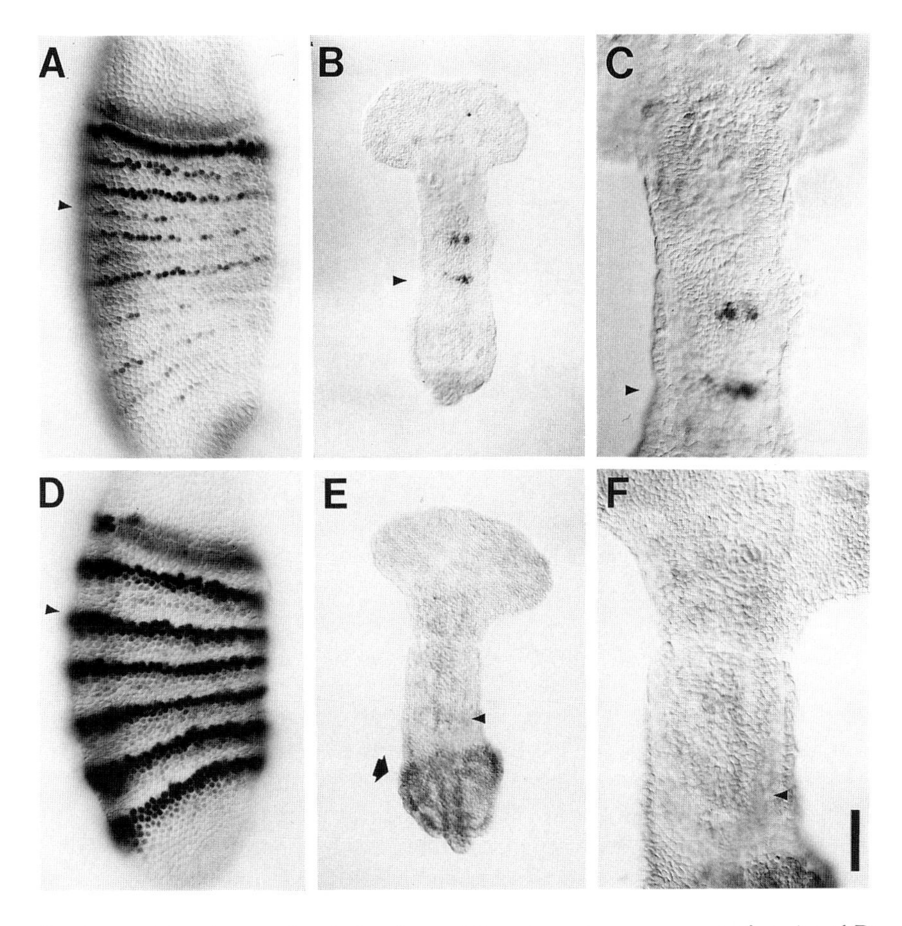

Fig. 6. *even-skipped* expression during *Drosophila* and grasshopper segmentation. **A** and **D**: *Drosophila engrailed* (A) and *even-skipped* (D) protein accumulation at the end of gastrulation shows that *even-skipped* stripes (D) are sharply defined at the time that *engrailed* stripes (A) are forming. Arrowhead marks the position of the posterior portion of the second thoracic (T2) segment (*engrailed* stripe 5, anterior part of *even-skipped* stripe 3). The anterior margin of each *even-skipped* stripe corresponds to each of the odd numbered *engrailed* stripes. Note the pair-rule pattern of stripe intensity in which the odd-numbered *engrailed* stripes are weaker then the even-numbered ones at this stage. **B** and **C**: At 17% of grasshopper development, *engrailed* stripes (B and a higher magnification view in C) have just appeared in T1 and T2 (arrowhead marks T2). The next stripes to appear will be further anterior in S1 and S3 (the mandibular and labial segments). The stripes of S1 and S3 represent the only pair of grasshopper *engrailed* stripes to appear in any pattern reminiscent of the pair-rule pattern of initiation seen for *Drosophila engrailed*. **E** and **F**: Grasshopper embryos of the same stage which have been stained for *even-skipped* expression; no pair-rule stripe pattern is visible. The position of T2, which can be determined even in unstained preparations by bulges in the mesoderm that are visible shortly after *engrailed* stripes appear, is marked by an arrowhead. The higher magnification view (in panel F) shows that no stripe of *even-skipped* expression is evident in T2, where *engrailed* expression has started, or more anteriorly in S1 or S3, where *engrailed* stripes will next appear. There is, however, a domain of grasshopper *even-skipped* expression in the more posterior mesoderm (arrow in E). At no earlier or later stage, do we observe *even-skipped* expression in a pattern of ectodermal stripes that would be consistent with a role in the regulation of *engrailed* expression. Scale bar = 200 μm (B and E); 100 μm (C and F); 75 μm (A and D). Anterior is up in all panels (from Patel et al., 1992).

establish the proper pattern of *engrailed* expression during *Drosophila* development.

To determine whether *even-skipped* might regulate *engrailed* in grasshopper, we examined the expression of *even-skipped* protein at stages prior to (12% to 16% of development) and during (17% to 30% of development) the initiation of *engrailed* stripe expression. Figures 6B and 6C show a grasshopper embryo in which the first two *engrailed* stripes (T1 and T2) have just started to appear. At this stage, *eve* (Fig. 6E, F) does not show any pair-rule pattern of expression, and eve expression does not overlap with that of *engrailed*. Later in development, when the *engrailed* stripes are appearing in the abdominal region, there is also no detectable pair-rule pattern of *even-skipped* expression. At no time in development do we observe an expression pattern that would suggest that grasshopper *eve* is involved in the initiation of *engrailed* stripe expression. In fact, with the exception of a few neurons, we have not observed grasshopper *eve* and *engrailed* localization within the same nucleus.

There are eight known pair-rule genes in Drosophila, but *even-skipped, hairy,* and *runt* are considered to be the three primary pair-rule genes (Small and Levine, 1991). These primary pair-rule genes are responsible for generating periodic patterns from the regional patterns of gap gene expression. The interactions of the three primary pair-rule genes regulate the expression of the remaining pair-rule genes (reviewed in Small and Levine, 1991). If pair-rule patterning were involved in short germband development, one might expect to find grasshopper homologs of at least some of the *Drosophila* primary pair-rule genes expressed in a pair-rule fashion prior to the appearance of *engrailed* stripes. Grasshopper *even-skipped,* however, does not display a pair-rule pattern of expression and therefore does not appear to play a role in setting up the pattern of *engrailed* expression during grasshopper development. These observations strengthen the argument that pair-rule patterning is not involved in the segmentation of short germband embryos. One caveat to this analysis is the potential existence of a second *eve* gene in grasshopper that does play a role in pair-rule patterning, but we found no indication of such a second gene.

Despite the lack of pair-rule expression, *eve* is expressed in the early grasshopper embryo. At the onset of gastrulation, *eve* protein accumulates in the posterior region of the embryo, mainly in the mesoderm, with low levels in the overlying ectoderm. At no point does this pattern resolve into a pattern of stripes. As the embryo begins to elongate, expression is maintained in the most posterior region of mesoderm (Fig. 3E), with the highest levels of expression in mesodermal cells in medial regions of the gastral groove. At about 25% of development, this mesodermal staining disappears. Thus, early expression in grasshopper seems to be limited to posterior mesoderm during the time that the embryo is gastrulating, an expression pattern unlike anything reported for *Drosophila eve.*

This expression in posterior mesoderm is, however, reminiscent of the early expression patterns of *Xhox-3* and *Evx-1, even-skipped* homologs in *Xenopus* and mouse respectively (Ruiz i Altaba and Melton, 1989a; Bastian and Gruss, 1990; Dush and Martin, 1992). These vertebrate genes are expressed predominately in the posterior mesoderm, and a variety of manipulations in *Xenopus* suggest that these genes may be involved in early anterior-posterior axial patterning, although the changes in the anterior-posterior axis may actually be the result of even earlier perturbations in dorsal-ventral mesodermal patterning (Ruiz i Altaba and Melton, 1989b; Dush and Martin, 1992). Later in development, the vertebrate *even-skipped* homologs are expressed within the central nervous system in a subset of spinal cord neurons. Interestingly, the vertebrate *even-skipped* homologs appear to be associated with HOX gene clusters. For example, the human *even-skipped* homologs, *EVX-1* and *EVX-2*, are located at the extreme 5′ ends of the clusters on chromosomes 7 and 2 respectively (Boncinelli et al., 1991). In general, the domain of expression of a HOX gene along the anterior-posterior axis corresponds to its position within the HOX cluster. Those at the 5′ end tend to be expressed in more posterior regions of the vertebrate embryo, and successively more anterior domains of expression are seen for each gene located more 3′ (reviewed in McGinnis and Krumlauf, 1992). Thus, the posterior domain of expression seen for vertebrate *eve* genes is consistent with their location at the 5′ end of the complexes. In *Drosophila,* however, *even-skipped* is located on a different chromosome from the one containing the *Antennapedia* and *Bithorax* homeotic complexes. It is not yet known whether the grasshopper *even-skipped* gene is located near any of the homeotic gene homologs.

The expression patterns of grasshopper *even-skipped* suggest that *eve* played a role in neurogenesis and/or axial patterning in the common ancestor to vertebrates and arthropods. The nearly identical expression patterns in neurons, in dorsal mesoderm, and in the ring of tissue at the anal pad suggest a highly conserved role for this protein in the germband stage of all insects. During the course of insect evolution, insects utilizing a long germband mode of development arose, and *even-skipped* acquired an additional function—that of pair-rule patterning.

The data provided by the analysis of *abdominal-A, engrailed,* and *even-skipped* expression supports some of the general conclusions reached from the previous embryological studies. The striking conservation of the germband stage in all arthropods suggested that genes active at this time of development would be highly conserved as well (Sander, 1988). All three genes do, indeed, show very similar patterns of expression during the germband stage of both *Drosophila* and grasshopper. The same embryological studies, however, indicated that gene interactions leading up to the germband stage might differ significantly between short and long germband embryos. This conclusion is supported by the initiation patterns of *engrailed* and *abdominal-A* expression and by the more recent discovery that *even-*

skipped does not appear to play a role in pair-rule patterning in grasshopper. These results provide strong evidence that some of the pattern formation steps seen in *Drosophila* do not apply to other insects. Certain aspects of Drosophila pattern formation may instead represent recent evolutionary advances and may differ from the mode of pattern formation used in the ancestors of present-day arthropods.

FUTURE DIRECTIONS

The various results obtained so far suggest that the expression of genes utilized at the germband stage (homeotic and segment polarity genes) will be relatively well conserved in all insects but that the genes acting earlier (maternal, gap, and pair-rule genes) may not have conserved developmental functions in pattern formation. To test the validity of this view, and to better understand the evolution of insect pattern formation, it will be important to isolate additional grasshopper homologs of *Drosophila* segmentation and homeotic genes.

Grasshopper homologs of the Drosophila homeotic genes *Sex combs reduced* and *Antennapedia,* in addition to *abdominal*-A, have been isolated and the distribution of their transcripts analyzed (Akam et al., 1988; E. Ball, N. Patel, and C. Goodman, unpublished data). In general, the domains of expression are similar to those seen in *Drosophila.* Detailed analyses of the expression of these genes in grasshoppers and other insects such as the beetles (*Tribolium*, see Chapter 6, this volume), however, may tell us about certain aspects of the evolution of the homeotic complexes during insect evolution. Specifically, evolutionary changes in the body plan pattern, such as the specialization of terminal abdominal segments, may be the result of subtle changes in the expression patterns of the homeotic genes. Such a correlation between terminal segment differentiation and *abdominal*-A expression has already been observed (Tear et al., 1990).

The isolation of additional segment polarity genes may also provide valuable information. For many genes of this class, especially those that also control neuroblast patterning (Patel et. al., 1989c), germband stage expression would be expected to be roughly similar in *Drosophila* and grasshopper. The final pattern of cuticular structures is different in various insects and may depend on changes in the expression patterns of segment polarity genes late in embryogenesis. In addition, data obtained from other insects may help us to understand the complex cellular interactions that regulate segment polarity gene expression in the later stages of *Drosophila* development (reviewed in Ingham and Martinez-Arias, 1992). There are already some indications that differences in later regulation might exist. In both *Drosophila* and grasshopper, the initial anterior boundary of the *engrailed* stripes corresponds to the parasegment boundary. Later in the development of both insects, the posterior border of *engrailed* expression appears to be aligned with segment boundary. In both insects, the alignment to the segment border occurs after the initial *engrailed* pattern has undergone some modification. In grasshopper, each initial *engrailed* stripe expands posteriorly, as formerly *engrailed*-negative cells begin to express *engrailed* (Patel et al., 1989b). By contrast, however, *Drosophila engrailed* stripes appear to narrow as previously engrailed-positive cells (or their progeny) cease expressing *engrailed* (Vincent and O'Farrell, 1992). In *Drosophila* embryos, this modification of *engrailed* expression occurs at a time when cells are changing neighbors due to the cell rearrangements that occur during germband extension. By contrast, grasshopper embryos do not undergo a similar phase of cell rearrangements. Since cell–cell interactions are known to be responsible for maintaining *engrailed* expression, these differences in cell rearrangements might result in the contrasting modifications of the *engrailed* pattern. This explanation is also consistent with the observations of *engrailed* stripe expansion that occurs in the honeybee embryo, a long germband embryo that does not undergo germband extension (Fleig, 1990).

One additional feature of grasshopper development that might prove useful in developmental studies arises from the fact that it is a hemimetabolous insect. That is, the grasshopper embryo develops adult structures such as legs and antennae directly. In *Drosophila,* a holometabolous insect, such adult structures develop postembryonically and are derived from imaginal discs. Many of the segment polarity and homeotic mutations have interesting effects on adult structures, and the analysis of the expression patterns of these genes in grasshopper embryos will help link embryonic pattern formation events to the development of adult structures.

The isolation and analysis of grasshopper homologs of the *Drosophila* pair-rule genes *hairy* and *runt* should clarify whether or not pair-rule patterning is involved in short germband development. Both of these genes, like *eve,* also serve functions during the later stages of *Drosophila* development. For example, *runt* is expressed in the developing nervous system and is required for the proper development of a subset of neurons (Kania et al., 1990; Duffy et al., 1991). Thus, homologs of these genes should exist in grasshopper for these later functions. Once the genes are isolated, we can ask then whether they show any pair-rule patterns of expression. Given that individual stripes of primary pair-rule genes expression are controlled by separable promoter elements (reviewed in Small and Levine, 1991), it is possible to think that *hairy* and/or *runt* stripes could appear one at a time, ahead of the forming *engrailed* stripes, during grasshopper development. On the other hand, if no pair-rule patterns are seen for either *hairy* or *runt* in grasshopper embryos, then, taken together with what we already have seen for *even-skipped,* it becomes very unlikely that pair-rule patterning is involved in segmentation in short germband embryos.

If pair-rule patterning does not establish the expression of grasshopper *engrailed,* then what does? One possibility is that the cell–cell interactions mediated by segment polarity genes that maintain *engrailed* expression in *Drosophila* are used to sequentially initiate *engrailed* stripes in grasshopper (reviewed in Ingham and Martinez-Arias, 1992). These cell–cell interactions might be begin in the T1/T2 region and spread anteriorly and posteriorly. A second possibility arises from the observation that *engrailed* stripes in the *Drosophila* head are formed and maintained by a mechanism that appears to be independent of at least certain pair-rule genes. This mechanism may represent a primitive mode of pattern formation that is utilized throughout short germband embryos. In *Drosophila,* genes with "gaplike" expression patterns, such as *orthodenticle, buttonhead,* and *empty spiracles,* appear to control pattern formation in the head (reviewed in St Johnston and Nüsslein-Volhard, 1992). Thus it may be that grasshopper segment polarity gene expression is directly controlled by a series of gap genes, bypassing genes of the pair-rule type.

The isolation of grasshopper homologs of *Drosophila* gap genes is another important goal. As with many segmentation genes, gap genes are expressed during later phases of *Drosophila* development; thus, it is reasonable to expect homologs to be present in grasshopper. The analysis of gap gene expression during early grasshopper development may suggest what roles, if any, these genes serve during short germband embryo development. If they are expressed regionally at the blastoderm stage, this would imply that the blastoderm of short germband embryos does contain some patterning information. An analysis of gap gene expression patterns might support the hypothesis that these genes regulate segment polarity and homeotic gene expression in grasshopper. The RNA and protein distribution patterns of gap genes, especially the *hunchback* gene, may also reveal the existence of gradients of positional information within the developing grasshopper egg.

What about the earliest steps of pattern formation, especially those that are associated with maternal pathways in Drosophila? Given the absence of nurse cells, it is unlikely that the ovary provides the oocyte with RNA encoded by potential homologs of such genes as *bicoid* and *nanos* in the grasshopper. One could imagine, however, that the external signalling mechanisms are more or less similar. In *Drosophila,* certain early steps of terminal and dorsal-ventral patterning are regulated by the follicle cells that surround the developing oocyte. Clear differences in follicle cell morphology, differentiation, and gene expression are evident along both the anterior-posterior and dorsal-ventral axes. The follicle cells appear to provide some positional information to the oocyte (reviewed in St Johnston and Nússlein-Volhard, 1992). For example, follicle cells contacting the anterior and posterior ends of the oocyte appear to send a signal to the oocyte that defines the terminal domains. Genetic analysis suggests that the *torso-like* gene is required in the terminal follicle cells to produce the localized terminal signal, which is deposited in the vitelline membrane surrounding the oocyte. After fertilization, the apparent receptor of this signal, the transmembrane receptor tyrosine kinase encoded by the *torso* gene, is uniformly distributed on the egg membrane. The localized terminal ligand then interacts with the receptor to provide the developing embryo with positional information defining the terminal regions of the embryo. Dorsal-ventral signalling also involves the interaction of the follicle cells with the developing oocyte, but in this case a continuous gradient of positional information is provided to the developing embryo.

It is possible that similar events could occur during grasshopper development. The imprints left by the follicle cells on the grasshopper chorion indicate that specialized follicle cells contact the posterior region of the egg where the embryo will eventually form. These follicle cells may provide external cues for terminal and dorsal-ventral patterning in grasshopper as they do in *Drosophila.* In addition, they could also influence anterior-posterior patterning in grasshopper. For example, it is possible that *bicoid* and *nanos* RNA are produced by the nuclei of the oocyte and/or early zygote and that they become localized to specific points at the surface of the egg due to the action of external cues from the ovary. In *Drosophila, nanos* RNA is transported to the future posterior pole, and it is not unreasonable to assume that this could also occur in grasshopper.

The discovery that *nanos* is not required when maternal *hunchback* is also eliminated suggests that there may be some redundancy in the anterior-posterior patterning mechanism (Hülskamp et al., 1989; Irish et al., 1989; Struhl, 1989). Embryological studies suggest that the posterior system may be evolutionarily more ancient (French, 1988); thus, grasshopper pattern formation may utilize a posterior determinant only. Finally, grasshopper anterior-posterior patterning could be more like *Drosophila* dorsal-ventral patterning. Rather than involving gradients within the egg during the syncytial stages, it may rely on gradients in the extracellular space outside the egg that contribute to embryonic pattern formation via a signal transduction pathway (St Johnston and Nüsslein-Volhard, 1992).

Although the limited molecular studies of grasshopper pattern formation support most of the embryological data, some of the interpretations of the embryonic manipulations may need to be reexamined. For example, short germband growth has been thought to involve a subterminal growth zone. During the development of the grasshopper abdomen, however, there is no indication of a localized region of mitosis during the entire phase of elongation (G. Tear, personal communication; Patel and Goodman, unpublished results). In addition, the developing *engrailed* stripes do not maintain a constant distance from the end of the germband. It appears instead that widespread cell proliferation causes the entire abdominal region to expand, and the sheet of cells is then patterned progressively from anterior to posterior as revealed by the sequential appearance of *engrailed* stripes. In a sense, this aspect of pattern formation is more similar to that of *Drosophila* than is implied by previous embryological descriptions, although it is important to remember that this patterning step in grasshopper takes place in a cellular, as opposed to a syncytial, environment. Furthermore, embryological studies had suggested that the head of short germband embryos is the first region to be determined. If we take the order of appearance of *engrailed* stripes to reflect the order of regional specification, however, then the grasshopper head region is not the first region to be specified. Rather, the thorax is first determined, and pattern formation then spreads both anteriorly and posteriorly. This is more consistent with Seidel's concept of a differentiation center than with the strict anterior to posterior progression of patterning suggested for short germband embryos (Sander, 1976). More accurate fate maps of short germband embryos will help in interpreting molecular results. In previous studies, localized damage was used to create fate maps, but more accurate interpretations can now be made using single-cell marking techniques.

In addition to the molecular studies of grasshopper development, similar studies in other insects will be required to obtain a complete picture of the evolution of insect segmentation. Several homeotic and segmentation genes have been cloned from *Apis* (bee), which, like *Drosophila,* utilizes the long germband mode of development (Waldorf et al., 1989). The development of insect embryos of the intermediate germband type, such as those of moths (*Manduca* and *Bombyx*) and beetles (*Tribolium*), has received considerable attention at both the molecular and genetic levels. These intermediate germband embryos develop in eggs that form in meroistic ovaries, and embryological manipulations clearly indicate the existence of gradients of positional information within the egg (French, 1988). In *Manduca* embryos, there is clear evidence for pair-rule patterning (Carr and Taghert, 1989; N.P. and C. Goodman, unpublished data), and *abdominal*-A expression appears to be initiated simultaneously in the entire abdominal domain as in *Drosophila* (Nagy et al., 1991). Careful observations of living moth embryos indicate that the "growth" that occurs in the transition from the blastoderm stage to the germband stage is predominantly due to cell rearrangements as opposed to a localized growth zone (L. Nagy, personal communication). Thus, intermediate germband development may be closely related to long germband development. Homologs of *Drosophila* gap, pair-rule, and homeotic genes have been isolated from *Tribolium* (see Chapter 6, this volume; R. Dennel, personal communication; R. Sommer and D. Tautz, personal communication), and the examination of their expression patterns will help establish the relationship between the intermediate and long germband modes of development.

Additional information regarding the evolution of segmentation will also come from an analysis of crustacean segmentation. The development of crustacean embryos, such as those of crayfish, is clearly comparable to insect development (Patel et al., 1989b). In some regards, they fit the definition of short germband development better than any insect embryo because the entire region posterior to the first thoracic segment develops from a well-defined growth zone. The germband is formed by an invariant lineage pattern (Dohle and Scholtz, 1988); thus, these embryos will be particularly useful for studying the cell–cell interactions that are mediated by segment polarity genes.

The molecular studies of grasshopper segmentation provide some initial answers to longstanding questions concerning the evolution of insect segmentation and, moreover, add interesting insights into the evolution of developmental systems. Future studies will aim not only to isolate additional homologs of *Drosophila* segmentation genes in various insects but also to explore ways to manipulate the expression of these genes in order to more clearly determine their roles in the development of these insects. It may also be possible to conduct genetic screens in other insects or arthropods to investigate more carefully the mechanisms of pattern formation in insects other than *Drosophila.* These types of analysis will provide valuable information regarding the evolution of developmental systems, and they should be broadly applicable to understanding the evolution of development within other phyla and between distantly related phyla.

ACKNOWLEDGMENTS

Many of the studies presented in this review were conducted in the lab of Dr. Corey Goodman, and his support and encouragement are gratefully acknowledged. In addition, the results described for grasshopper *eve* are the result of a fruitful collaboration with Dr. Eldon Ball. I would also like to thank Guy Tear, Ruth Lehmann, Allan Spradling, Ralf Sommer, Rob Dennel, David Bentley, and Lisa Nagy for helpful discussions, and Becky Chasan for comments on the manuscript.

REFERENCES

Akam M (1987): The molecular basis for metameric pattern in the *Drosophila* embryo. Development 101:1–22.

Akam M, Dawson I, Tear G (1988): Homeotic genes and the control of segment diversity. Development 104 (Suppl.):123–134.

Akam M (1989): *Hox* and HOM: homologous gene clusters in insects and vertebrates. Cell 57:347–349.

Bastian H, Gruss P (1990): A murine *even-skipped* homologue, *Evx 1*, is expressed during early embryogenesis and neurogenesis in a biphasic manner. EMBO J 9:1839–1852.

Boncinelli E, Simeone A, Acampora D, Mivilio F (1991): HOX gene activation by retinoic acid. Trends Genet 7:329–334.

Carr JN, Taghert PH (1989): Pair-rule expression of a cell-surface molecule during gastrulation of the moth embryo. Development 107:143–151.

DiNardo S, Kuner JM Theis J, O'Farrel PH (1985): Development of embryonic pattern as revealed by accumulation of the nuclear *engrailed* protein. Cell 43:59–69.

DiNardo S, O'Farrell PH (1987): Establishment and refinement of segmental pattern in the *Drosophila* embryo: Spatial control of *engrailed* expression by pair-rule genes. Genes Dev 1:1212–1225.

Doe CQ, Smouse D, Goodman CS (1988): Control of neuronal fate by the Drosophila segmentation gene *even-skipped.* Nature 333:376–378.

Dohle W, Scholtz G (1988): Clonal analysis of the crustacean segment: The discordance between genealogical and segmental borders. Development 104(Suppl.):147–160.

Dressler GR, Deutsch U, Balling R, Simon D, Guenet JL, Gruss P (1988): Murine genes with homology to *Drosophila* segmentation genes. Development 104:(Suppl.):181–186.

Driever W, Nüsslein-Volhard C (1988): A gradient of *bicoid* protein in *Drosophila* embryos. Cell 54:83–93.

Duffy JB, Kania MA, Gergen JP (1991): Expression and function of the *Drosophila* gene *runt* in early stages of neural development. Development 113:1223–1230.

Dush M, Martin G (1992): Analysis of mouse *Evx* genes:*Evx-1* displays graded expression in the primitive streak. Dev Biol 151 (in press).

Field KG, Olsen G, Lane D, Giovannoni SJ, Ghiselin MT, Raff EC, Pace NR, Raff RA,(1988): Molecular phylogeny of the animal kingdom. Science 239:748–753.

Fjöse A, McGinnis W, Gehring WJ (1985): Isolation of a homeobox-containing gene from the *engrailed* region of *Drosophila* and the spatial distribution of its transcript. Nature 313: 284–289.

Fleig R (1990): *Engrailed* expression and body segmentation in the honeybee *Apis millifera*. Roux's Arch Devl Biol 198: 467–473.

Frasch M, Hoey T, Rushlow C, Doyle H, Levine M (1987): Characterization and localization of the *even-skipped* protein of *Drosophila*. EMBO J 6:749–759.

Frasch M, Warrior R, Tugwood J, Levine M (1988): Molecular analysis of even-skipped mutants in *Drosophila* development. Genes Dev 2:1824–1838.

French V (1988): Gradients and insect segmentation. Development 104(Suppl.):3–16.

Frigerio G, Burri M, Bopp D, Baumgartner S, Noll M (1986): Structure of the segmentation gene *paired* and the *Drosophila* PRD gene set as part of a gene network. Cell 47:735–746.

Graham A, Papalopulu N, Krumlauf R (1989): The murine and *Drosophila* homeobox gene complexes have common features of organization and expression. Cell 57:367–378.

Hülskamp M, Schröder, C Pfeifle C, Jäckle H, Tautz D (1989): Posterior segmentation of the *Drosophila* embryo in the absence of a maternal posterior organizer gene. Nature 338:629–632.

Ingham P (1988): The molecular genetics of embryonic pattern formation in *Drosophila*. Nature 335:25–34.

Ingham P (1991): Segment polarity genes and cell patterning within the *Drosphila* segment. Curr Opinions Genet Dev 1:261–267.

Ingham P, Martinez-Arias A (1992): Boundaries and fields in early embryos. Cell 68:221–236.

Irish V, Lehmann R, Akam M (1989): The *Drosophila* posterior-group gene *nanos* functions by repressing hunchback activity. Nature 338:646–648.

Joyner AL, Martin G (1987): *En-1* and *En-2*, two mouse genes with sequence homology to the *Drosophila engrailed* gene: Expression during embryogenesis. Genes Dev 1:29–38.

Jurgens G, Wieschaus E, Nüsslein-Volhard C, Kluding H (1984): Mutations affecting the pattern of the larval cuticle in *Drosophila melanogaster*. II. Zygotic loci on the third chromosome. Roux's Arch Dev Biol 193:283–295.

Kamb A, Weir M, Rudy B, Varmus H, Kenyon C (1989): Identification of genes from pattern formation, tyrosine kinase, and potassium channel families by DNA amplification. Proc Natl Acad Sci USA 86:4372–4376.

Kania MA, Bonner AS, Duffy JB, Gergen JP (1990): The *Drosophila* segmentation gene *runt* encodes a novel nuclear regulatory protein that is also expressed in the developing nervous system. Genes Dev 4:1701–1713.

King RC, Büning J (1985): The origin and functioning of insect oocytes and nurse cells. In Kerkut GA, Gilbert LI (eds): "Comprehensive Insect Physiology, Biochemistry, and Pharmacology." New York: Pergamon Press, pp 37–82.

Kornberg T, Siden I, O'Farrell PH, Simon M (1985): The *engrailed* locus of *Drosophila*: in situ localization of transcripts reveals compartment-specific expression. Cell 40: 40–53.

Krause G (1939): Die Eitypen der Insekten. Bio Zbl 59:495–536.

Lawrence PA (1988): The present status of the parasegment. Development 104 (Suppl.):61–66.

Lawrence PA, Johnston P, Macdonald P, Struhl G (1987): Borders of parasegments in *Drosophila* embryos are delimited by *fushi tarazu* and *even-skipped* genes. Nature 328:440–442.
Lewis EB (1978): A gene complex controlling segmentation in *Drosophila*. Nature 276: 565–570.
Macdonald PM, Ingham P, Struhl G (1986): Isolation, structure, and expression of *even-skipped*: A second pair-rule gene of *Drosophila* containing a homeo box. Cell 47:721–734.
Manton SM (1977): "The Arthropoda: Habits, Functional Morphology, and Evolution." Oxford: Clarendon Press.
Moloo SK (1971): The degree of determination of the early embryo of *Schistocera gregaria* (Forskål) (Orthoptera: Acrididae). J Embryol Exp Morph 25:277–299.
McGinnis W, Krumlauf R (1992): Homeobox genes and axial patterning. Cell 68:283–302.
Mee J (1986): Pattern formation in fragmented eggs of the short germ insect, *Schistocerca gregaria*. Roux's Arch Dev Biol 195:506–512.
Mee J. French V (1986): Disruption of segmentation in a short germ insect embryo. I. The localization of segmental abnormalities induced by heat shock. J Embryol Exp Morph 96:245–266.
Nagy LM, Booker R, Riddiford LM (1991): Isolation and embryonic expression of a *abdominal-A*-like gene from the lepidopteran, *Manduca sexta*. Development 112:119–129.
Nüsslein-Volhard C, Wieschaus E (1980): Mutations affecting segment number and polarity in *Drosophila*. Nature 287:795–801.
Nüsslein-Volhard C Wieschaus E, Kluding H (1984): Mutations affecting the pattern of the larval cuticle in *Drosophila melanogaster*. I. Zygotic loci on the second chromosome. Roux's Arch Dev Biol 193:267–282.
Patel NH, Kornberg TB, Goodman CS (1989a): Expression of *engrailed* during segmentation in grasshopper and crayfish. Development 107:201–212.
Patel NH, Martin-Blanco E, Coleman KG, Poole SJ, Ellis MC, Kornberg TB, Goodman CS (1989b): Expression of *engrailed* proteins in arthropods, annelids, and chordates. Cell 58:955–968.
Patel NH, Schafer B, Goodman CS, Holmgren R (1989c): The role of segment polarity genes during *Drosophila* neurogenesis. Genes Dev 3:890–904.
Patel NH, Ball EE, Goodman CS (1992): The changing role of *even-skipped* during the evolution of insect pattern formation. Nature (in press).
Peifer M, Karch F, Bender W (1987): The bithorax complex: Control of segment identity. Genes Dev 1:891–898.
Roonwal ML (1936): Studies on the embryology of the African migratory locust *Locusta migratoria migratoroides*. I. Early development with a new theory of multiphased gastrulation among insects. Phil Trans R Soc B 226:391–421.
Ruiz i Altaba A, Melton DA (1989a): Bimodal and graded expression of the Xenopus homeobox gene *Xhox3* during embryonic development. Development 106:173–183.
Ruiz i Altaba A, Melton D (1989b): Involvement of the Xenopus homeobox gene *Xhox3* in pattern formation along the anterior-posterior axis. Cell 57:317–326.
Sander K (1976): Specification of the basic body plattern in insect embryogenesis. Adv Insect Physiol 12:125–238.
Sander K (1983): The evolution of patterning mechanisms: Gleanings from insect embryogenesis and spermatogenesis. In Goodwin BC, Holder N, Wylie CC (eds): "Development and Evolution." Cambridge, England: Cambridge University Press, pp 137–159.
Sander K (1988): Studies in insect segmentation: From teratology to phenogenetics. Development 104 (Suppl.):112–121.
Schwalm FE (1988): Insect morphogenesis. In Sauer HW (ed): "Monographs in Developmental Biology." Basel: Karger Press.
Small S, Levine M (1991): The initiation of pair-rule stripes in the *Drosophila* blastoderm. Curr Opins Genet Dev 1:255–260.
St Johnston D, Nüsslein-Volhard C (1992): The origin of pattern and polarity in the *Drosophila* embryo. Cell 68:201–220.

Struhl G (1989): Differing strategies for organizing anterior and posterior body pattern in *Drosophila* embryos. Nature 338:741–744.

Tear G, Akam M, Martinez-Arias A (1990): Isolation of an *abdominal-A* gene from the locust *Schistocerca gregaria* and its expression during embryogenesis. Development 110:915–925.

Tear G, Bate CM, Martinez-Arias A (1988): A phylogenetic interpretation of the patterns of gene expression in *Drosophila* embryros. Development 104 (Suppl.):135–146.

Valentine JW (1989): Bilaterans of the Precambrian-Cambrian transition and the annelid-arthropod relationship. Proc Natl Acad Sci USA 86:2272–2275.

Vincent JP, O'Farrell PH (1992): The state of *engrailed* expression is not clonally transmitted during early *Drosophila* development. Cell 69:923–931.

Waldorf U, Fleig R, Gehring W (1989): Comparisons of the homeobox-containing genes of the honeybee and *Drosophila.* Proe Natl Acad Sci USA 86:9971–9975.

Wang C, Lehmann R (1991): *Nanos* is the localized posterior determinant in *Drosophila.* Cell 66:637–647.

Wieschaus E, Nüsslein-Volhard C, Jurgens G (1984): Mutations affecting the pattern of the larval cuticle in *Drosophila melanogaster.* III. Zygotic loci on the X chromosome and the fourth chromosome. Roux's Arch Dev Biol 193:296–307.

Evolutionary Conservation of Developmental Mechanisms, pages 111–123

8. Homeobox Genes in Plant Development: Mutational and Molecular Analysis

Erik Vollbrecht, Randy Kerstetter, Brenda Lowe, Bruce Veit, and Sarah Hake

USDA-ARS Plant Gene Expression Center, Albany, California 94710 (E.V., B.L., S.H.) Department of Plant Biology, University of California, Berkeley, California 94720 (R.K., B.V., S.H.)

INTRODUCTION

Animals and higher plants display fundamental differences in their developmental strategies (Walbot, 1983; Goldberg, 1988). In animals, cell movements are a foundation of numerous developmental programs. Major organ and tissue systems, or the rudiments thereof, form during embryogenesis. Postembryonic development is characterized by elaboration of adult structures. As animal development proceeds through successive stages, the developmental capacity of somatic cells becomes progressively reduced. In higher plants, on the other hand, cells are entirely nonmotile. During embryogenesis, polar root and shoot meristems are organized and only a few archetypes of adult vegetative structures are initiated. The majority of the plant body forms after germination as meristematic regions generate vegetative structures repetitively. Terminal differentiation of the shoot or axillary meristem results in the formation of reproductive floral structures (Sussex, 1989). Moreover, cell fate within developing organs is defined relatively late (Poethig, 1987; McDaniel and Poethig, 1988; Langdale et al., 1989), and even after final differentiation many plant cells retain great plasticity in developmental potential.

Despite the apparent contrasts between animal and plant developmental systems, both strategies require mechanisms that specify cell fate in proper spatial and temporal contexts. In plants, these mechanisms are still largely unknown. Genetic and biochemical studies on diverse animal systems suggest that control of cell fate, whether by autonomous or inductive processes (Davidson, 1990), involves relatively few types of genes (reviewed by Melton, 1991). One class of such genes consists of those encoding transcription factors. Transcription factors are utilized in specification of both animal and fungal cell types (Herskowitz, 1989; Wingender, 1990), and homeobox genes play an especially prominent role in the specification process (Gehring, 1987;

Scott et al., 1989; Holland and Hogan, 1988; Ruvkun and Finney, 1991). The homeobox is a nucleotide sequence, first found in homeotic genes of *Drosophila melanogaster,* that encodes a conserved sequence-specific DNA-binding motif called the homeodomain (McGinnis et al., 1984; Scott and Weiner, 1984). The role of transcription factors in programs that specify plant cell fate is becoming more apparent. Mutational analysis of the terminal meristem differentiation process has identified genes involved in floral development, some of which encode presumed transcription factors (Yanofsky et al., 1990; Sommer et al., 1990). Similarly, the *Knotted* gene (*Kn1*) of maize, which is defined by dominant mutations that alter the developmental fates of specific leaf cells (Bryan and Sass, 1941; Genlinas et al., 1969; Freeling and Hake, 1985), encodes a protein containing a homeodomain, the first identified in plants. The presence of a homeobox strongly suggests that *Kn1* also encodes a transcriptional regulator (Vollbrecht et al., 1991). Thus, regulation of gene expression by homeodomain proteins may represent a general mechanism of cell type specification that is broadly conserved in evolution.

With the discovery of the *Kn1* homeobox we consider how a single, evolutionarily conserved, molecular theme like the homeobox is deployed in fundamentally different developmental systems. As a comparative illustration of homeobox function and evolution, we describe analysis of *Kn1* and of additional homeobox genes isolated from maize and other plants.

MUTATIONS OF *KNOTTED* AFFECT LEAF DEVELOPMENT

The maize leaf consists of a blade and sheath, which are separated by a nonessential flap of tissue called the ligule (Fig. 1A). A hierarchical arrangement of leaf vasculature features a prominent midvein that forms in the center of the blade. Secondary lateral veins are spaced at intervals across the leaf, and subsidiary intermediate veins are between the laterals (Fig. 1A). Lateral and intermediate veins occur in both leaf blade and sheath. Development of the leaf can be broken down into three stages (Sylvester et al., 1990). The preligule stage, during which leaf shape is generated, consists of leaf initiation and widespread cell division. As is typical of plant organ development, the generation of shape is followed by a positional gradient of histogenesis. Thus, the second stage consists of ligule outgrowth and blade differentiation, and the final stage consists of sheath differentiation. Hence, vascular tissues of the blade and sheath, both of which are initiated in the preligule stage, differentiate during the second and third stages, respectively.

The *Knotted (Kn1)* locus is defined by a number of dominant neomorphic mutations that perturb maize leaf development (Freeling and Hake, 1985). Mutations at *Kn1* specifically affect the fate of cells along lateral veins in the leaf blade. Foci of cells frequently continue to grow relative to surrounding cells, resulting in pocketed outgrowths or knots (Fig. 1B). The

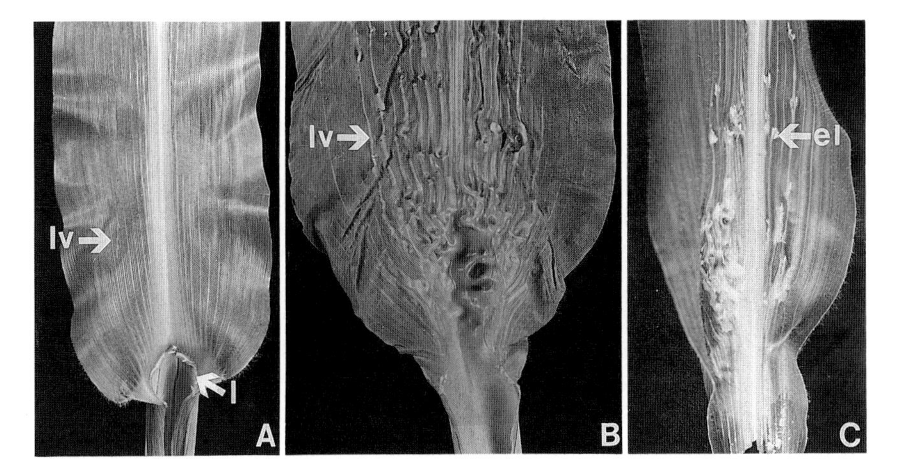

Fig. 1. Normal and knotted adult maize leaves. **A:** A normal leaf with the position of the ligule (1) indicated between the blade (above) and sheath (below). The position of a lateral vein (lv) is also indicated. **B:** A knotted leaf. The ligular region is highly disorganized, and lateral veins form outpocketings or knots, typical of mutant plants. **C:** This mutant leaf shows relatively few knots and substantial ectopic ligule (el) patches in the blade.

cells within and immediately surrounding knots show epidermal features and photosynthetic enzyme expression profiles (Sinha and Hake, 1991) more typical of cells normally found in the sheath. Thus the mutations confer a sheath-like fate on cells around blade lateral veins, but do not interfere with the spacing pattern of lateral vein placement. When lateral veins near the leaf base are strongly affected, the ligular region is often disrupted such that the blade–sheath boundary appears distorted, and sections of ligule are displaced into the blade (Figs. 1B, C). Ectopic ligule can also form de novo in the blade, along lateral veins that show additional features of the phenotype. Since an ectopic ligule borders on patches of affected (sheath-like) tissue, its appearance is consistent with its normal placement at a blade–sheath boundary and could be interpreted as a consequence of the juxtaposition of blade and sheath cell types. In accordance with the localization of the mutant phenotype to veins, and the fact that vascular tissues derive from the innermost of the leaf primordium's cell layers (Sharman, 1942), clonal analysis reveals that only the innermost cell layer around a lateral vein requires the mutant genotype for expression of the knotted phenotype (Hake and Freeling, 1986; Sinha and Hake, 1990). Since all cell layers are affected, the mutations must generate a signal that can specify cell identity in adjacent layers. When a sector boundary bisects a lateral vein, the entire mosaic vein displays the knotted phenotype, but adjacent veins in the wild type sector are unaffected. Thus, the signal is not transmitted transversely between lateral veins.

The phenotype associated with loss of *Kn1* function is not known. A small deletion that includes *Kn1* but none of the closest RFLP or genetic markers is lethal when hemizygous (uncovered by a translocation removing most of chromosome arm 1L), but the precise extent of the deletion is unknown (Veit et al., 1990). Several experiments are in progress to generate loss-of-function alleles from the dominant mutations. The dominant knotted phenotype could be interpreted as analogous to neomorphic mutations of *Drosophila* homeobox genes, for example the classic *Antennapedia* mutations (Schneuwly et al., 1987). In the context of this analogy, we could expect the loss-of-function phenotype to differ from that conditioned by the dominants, but it is difficult to envision a phenotype that reflects the bidirectional segment transformations characterizing mutations of homeotic genes like those in the ANT-C or BX-C complexes.

MOLECULAR ANALYSIS OF *KNOTTED*

Dominant mutations at *Kn1* affect noncoding regions of the gene. At least twelve mutations result from insertions into the third intron, while one mutation (*Kn1-O*) is a tandem duplication (Veit et al., 1990) (Fig. 2). The

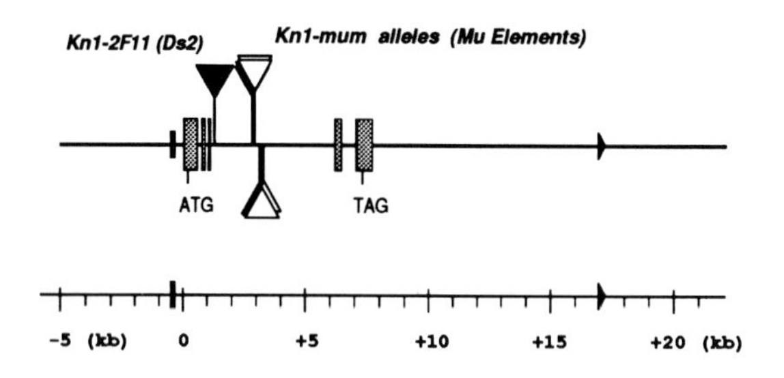

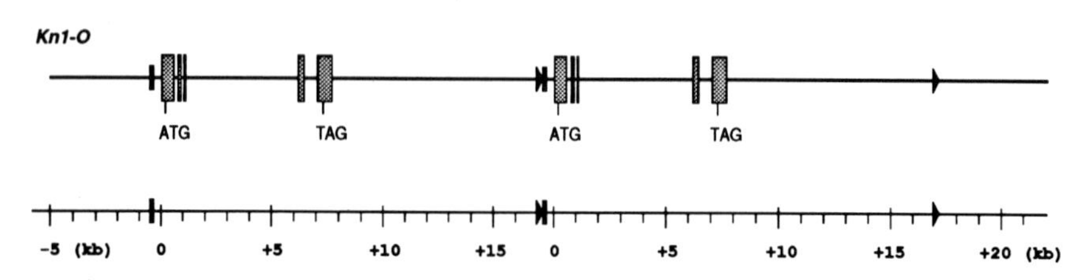

Fig. 2. The *Kn1* locus and structure of mutations. Exons are shown as shaded boxes, with the translation initiation and termination codons indicated. Zero on the coordinate maps is the transcription initiation site, and each unit represents 1.0 kb. **Upper**: Open triangles are insertions (not to scale) that cause various mutations; insertions map between +1.3 and +2.8. **Lower**: The *Kn1-0* tandem duplication. The extent of the *Kn1-0* repeat unit is denoted by a filled arrowhead and tail; the endpoints are at −0.45 and +17.0.

duplication results of a rearrangement in the 5' noncoding region of one transcription unit and leaves the other transcription unit in a normal context. Genetic and molecular analysis suggest that the cause of mutant phenotype in *Kn1-O* is the novel sequence upstream of one transcription unit, and not the simple dosage increase associated with a duplication (Freeling and Hake, 1985; Veit et al., 1990). The insertions in the third intron that we have characterized are all nonautonomous elements of the *Ac/Ds* or *Mutator* transposon families (Hake et al., 1989; R. Walko and S. H., unpublished results). In these alleles, the penetrance and expressivity of the knotted phenotype varies with the presence of the corresponding autonomous element. When transposase is present in *trans,* the penetrance and severity of the phenotype are greatly enhanced. Because the phenotype is still expressed in the absence of transposase, element excision is unlikely to play a role. Rather, we suggest that the physical presence of the transposase bound to the inserted element affects *Kn1* expression.

The *Kn1* gene encodes a single 1.7-kb mRNA. The message is detectable in all tissues examined to date, but is most abundant in rapidly dividing tissues such as inflorescence primordia (B. Greene and B. V., unpublished results). In both the insertion mutants and the tandem duplication, steady-state mRNA levels are not significantly altered. There appear to be no novel messages, suggesting that effects on RNA expression, if there are any, are subtle (Smith et al., 1992). One hypothesis supported by the collection of molecular and genetic data is that the mutations are regulatory and do not alter the protein coding capacity of the gene.

Initial results, using an antibody to the *Kn1* protein, support the hypothesis that the mutations affect regulation of the *Kn1* gene. In normal plants, the *Kn1* protein is detected in the nuclei of cells in vegetative and floral meristems but not in leaf primordia. Antibody staining is also detected in a subset of developing vascular bundles in the shoot apex region. In mutant plants, the protein shows similar meristem localizations, but the protein is also found sporadically in the developing lateral veins of leaf primordia (Smith et al., 1992). This finding corroborates the clonal analysis results, which showed that internal, vascular cells induce multiple cell layers to produce knots and ectopic ligules (Sinha and Hake, 1990). We propose a model which assumes that *Kn1* specifies a meristem or developing provascular identity to cells in which it is expressed. If ectopic *Kn1* expression occurs in cells of leaf primordia during the developmental stage when blade and ligular regions differentiate, these cells might be prevented from acquiring differentiated fates and be marked for additional (meristematic) growth relative to neighboring cells. This localized extra growth could explain the structural formation of knots. The subsequent removal of *Kn1* might then allow cells to respond to differentiation programs in progress (Freeling and Hake, 1985), which at this time would logically correspond to the developmental stage in which sheath tissues differentiate. The delay in histogenesis would account for the sheathlike qualities of cells in the blade. Ectopic ligules, as suggested, may form as a consequence of juxtaposing cells of blade and sheath identity.

PLANT HOMEOBOX GENES

The discovery of the *Kn1* homeobox facilitates the isolation of additional plant homeobox genes (Vollbrecht et al., 1991). Related homeobox genes have been found in all plant species examined, including *Arabidopsis* (K. Serikawa, S. Hake and P. Zambryski, unpublished results), rice (R. Walko, unpublished results) and tomato (N. Sinha and S. H., unpublished results). We have identified approximately fifteen different maize homeobox genes to date. The ten for which we have sequence information all encode homeodomains (HDs) that are quite similar to KN1 in primary amino acid sequence. Notably, a stretch of 13 amino acids that includes the putative recognition helix of the helix-turn-helix motif is completely conserved in all of the plant HDs we have isolated (Fig. 3). We assume that the plant HD proteins are not functionally redundant, that they participate in regulation of various cellular processes, and that they thus might recognize different target sequences. The absolute sequence conservation within the putative recognition helix implies that, as in classic HD proteins, multiple regions probably modulate DNA binding specificity (Kuziora and McGinnis, 1989; Gibson et al., 1990; Hayashi and Scott, 1990; Mann and Hogness, 1990; Damante and Di Lauro, 1991). An additional block of amino acids (ELK region, see below) that abuts the N-terminus of the HD (Vollbrecht et al., 1991) is also highly conserved in all of the plant HDs. Finally, each of the maize genes has an intron in an absolutely conserved position within the homeobox, and subgroups share the occurrence of up to two more introns in the adjacent ELK region (E.V. and R.K., unpublished results). The position of the intron interrupting the plant homeoboxes is distinct from the position of introns in animal or fungal homeoboxes (see Allen et al., 1991 for recent catalogue of homeobox intron positions).

We have mapped the cloned maize homeoboxes to chromosomal locations using recombinant inbred lines (Burr et al., 1988). Clusters of homeobox genes homologous to the genes of ANT-C and/or BX-C of *Drosophila* (Lewis, 1978) have been identified in an array of animals, including vertebrates (Graham et al., 1989), and the nonsegmented invertebrate *C. elegans* (Burglin et al., 1991; Kenyon and Wang, 1991). Nine maize loci have been placed on the RFLP map and we find no evidence of clustering (B. Lowe, unpublished results). The lack of clustering may reflect a fundamental evolutionary distinction between *Kn1*-like and *Antp*-like HDs (see below). Two of the mapped homeobox genes show tight genetic linkage to known mutations, namely the dominant morphological mutation *Rough sheath* (P. Be-

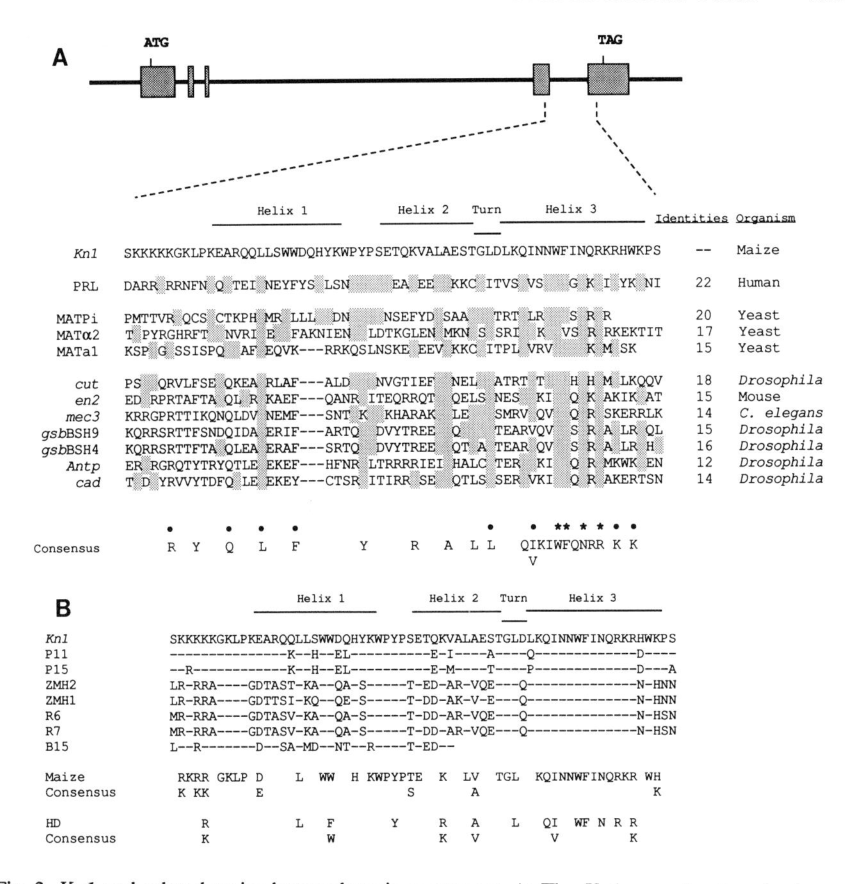

Fig. 3. *Kn1* and related maize homeodomain sequences. **A**: The *Kn1* gene structure is above, dotted lines indicating the region that encodes the homeodomain. The schematic Helix-Turn-Helix (HTH) framework is positioned according to the corresponding motif of the ANTP and engrailed homeodomains. Below, the amino acid sequence of several homeodomains is aligned with a segment of the KN1 sequence. The three-residue gap (hyphens in sequences) was introduced to maximize overall sequence homology. Positions with residues identical to those of KN1 are shaded. Stars identify the four residues invariant among multicellular eukaryotic homeodomains, and the eight most highly conserved positions are indicated by dots. The consensus (Scott et al., 1989) is derived from a collection of animal and fungal sequences (from Vollbrecht et al., 1991). **B**: An alignment of maize homeodomain sequences. Hyphens indicate residues identical to those of KN1. The sequence of clone B15 is incomplete. The maize consensus is derived from the sequences shown, and the homeodomain (HD) consensus is derived from the consensus in Figure 3A and the maize consensus.

craft, S.H., and M. Freeling, unpublished results) and the delayed greening mutation *argentia* (N. Sinha and S.H., unpublished results). Parsimony analysis of the maize sequences delimits three distinct subgroups, two of which have multiple members (Fig. 4A). Interestingly, amino acid sequences outside of the ELK and HD regions are conserved between members of distinct subgroups, suggesting closely related proteins. The intriguing possibility that

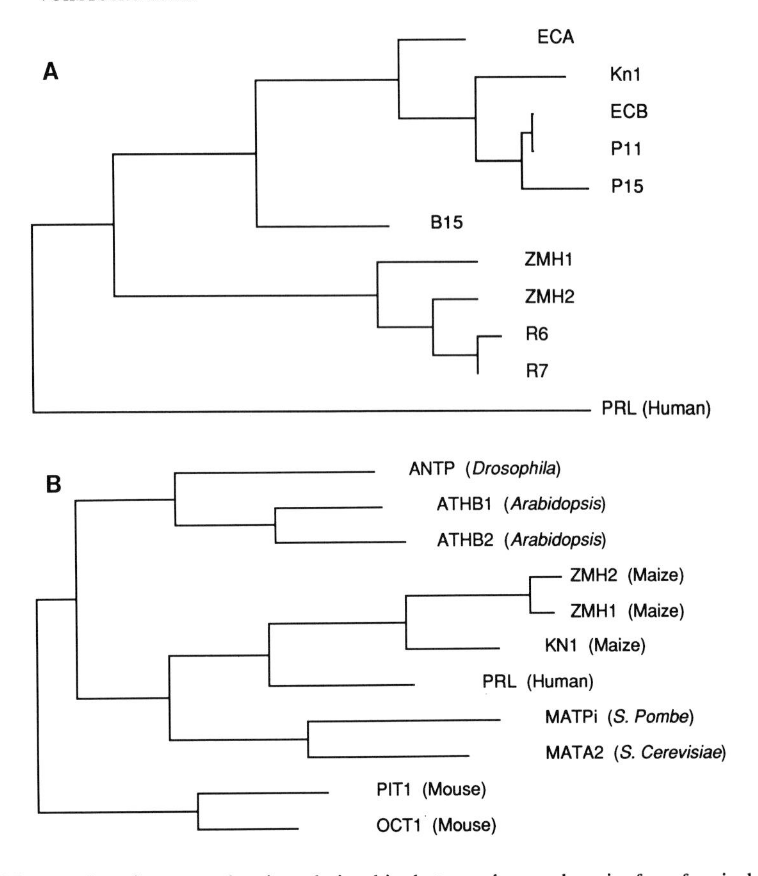

Fig. 4. Most parsimonious trees showing relationships between homeodomains from fungi, plants, and animals. **A**: Ten maize HDs can be broken down into two monophyletic groups. The tree was constructed using amino acid sequences extending from the ELK region (see Fig. 5) into the adjacent HD. PRL is used as an outgroup to root the tree. **B**: Parsimony analysis of homeodomain amino acid sequences indicates possible evolutionary relationships between two subgroups. The lower group defines the *Kn1*-like HDs (see text), and includes plant, animal, and fungal proteins. POU domain proteins (Herr et al., 1988) PIT1 and OCT1 are used as an outgroup to root the tree. The tree was generated using the three-residue gap indicated in Figure 3, but the same two monophyletic groups are obtained when the gap is removed.

the subgroups define genes that are expressed in similar tissue or cell types, or are involved in the regulation of like processes, is being investigated.

Characterization of homeobox genes in other plants could lead to a greater understanding of the action of individual genes like *Kn1*. For example, the *engrailed* protein is required in *Drosophila* development during embryogenesis and again later during neurogenesis. In other animals, however, *engrailed* is expressed only during neurogenesis, suggesting that its function in neurogenesis is ancient (Patel et al., 1989). Although data for the normal function of the *Kn1* gene product are just emerging, a similar approach might lead to an understanding of this plant gene's ancestral function. This approach would logically begin with a comparison of genes from evolutionarily distant plants, some of which may contain *Kn1* homologs.

The maize homeobox genes encode products that belong to an evolutionarily distinct subfamily of HD sequences, as illustrated by a comparison of the *Antp* and *Kn1* HDs (Fig. 3A). A three-residue gap must be inserted in the *Antp* sequence to achieve optimal similarity. The gap retains alignment at the homeodomain N-terminus of two highly conserved residues, and of presumably analagous clusters of basic amino acids. In *Antp*-like HDs the basic cluster contacts DNA, and the highly conserved Leu residue forms a part of the hydrophobic core of the folded motif (Qian et al., 1989; Kissinger et al., 1990). The subfamily of slightly larger (64 aa vs. 61 aa) HDs includes the human PRL protein and the *S. pombe* MATPi gene product (Fig. 4b). We will henceforth refer to homeodomains of this ancient subfamily as *Kn1*-like. Interestingly, PRL and MATPi are involved in specification of cell fate in their respective systems (Kamps et al., 1990; Nourse et al., 1990; Kelly et al., 1988), as is maize *Kn1*. Recently characterized HD proteins ATHB-1 and ATHB-2 from *Arabidopsis* (Ruberti et al., 1991) are not *Kn1*-like, but are *Antp*-like in size (61 aa) and amino acid sequence (Fig. 4B). This distinction suggests that at least two HD prototypes were present prior to the divergence of plants and animals. We offer two general proposals to account for the evolutionary retention of *Kn1*-like HDs in such diverse organisms. Perhaps ancient *Kn1*-like HD proteins had features that predisposed them to be recruited during evolution for assorted developmental functions. On the other hand, *Kn1*-like HDs may have been components of an ancient regulatory mechanism that was differentially assimilated into what became divergent developmental programs.

While HDs may be common to all eukaryotes, the *Kn1*-like class appears to be absent from *Drosophila*. Given the number of homeobox genes identified in *Drosophila*, including many homeobox loci first identified genetically (Scott et al., 1989), it is striking that none are *Kn1*-like. If the *Kn1*-like class is truly absent from *Drosophila,* it will undoubtedly be absent from many animals.

The plant *Kn1*-like homeoproteins all share an additional motif that occurs immediately on the N-terminal side of the homeodomain (Vollbrecht et al., 1991). We have named it the ELK region because of an invariant series of consecutive Glu, Leu, and Lys amino acids. The motif spans at least 24 amino acids, resulting in an extended conserved block (ELK plus HD) of 88 amino acids (Fig. 5). It is not present in animal or fungal HD proteins, nor in other proteins in the data bases. The ELK region is distinguished by highly conserved leucine or hydrophobic residues, interspersed in a region of predominantly charged and polar amino acids. The periodicity of the hydrophobic residues is distinct from that of leucine zippers (Landschulz et al., 1988) or previously described amphipathic alpha helices (Murre et al., 1989). The

```
                               ELK region                            Homeodomain
KN1   ... E V D A H G V D Q E L K H H L L K K Y S G Y L S S L K Q E L  S K K K K K G K L ...
ZMH1  ... R S L M E R V R Q E L K M E L K Q G F K S R I E D V R E E I  L R K R R A G K L ...
ZMH2  ... R S L V E R V R Q E L K H E L K Q G Y R D K L V D I R E E I  L R K R R A G K L ...
R6    ... R S L V E S V R K E L K N E L K Q G Y K E K L V D I R E E I  M R K R R A G K L ...
R7                ... V R K E L K N E L K Q G Y K E K L V D I R E E I  M R K R R A G K L ...
EC5   ... G I D P C S D D K E L K K Q L L R K Y S G C L G N L R K E L  C K K R K K G K L ...
EC21  ... E I D P R A E D K E L K Y Q L L K K Y S G Y L S S L R Q E F  S K K K K K G K L ...
P11                     ... E L K Y Q L L K K Y S G Y L S S L R Q E F  S K K K K K G K L ...
P15                     ... E L K H Q L L R K Y G G Y L G R L R Q E F  S K K K K K G K L ...
B151                    ... E L K E M L L K K Y S G C L S R L R S E F  L K K R K K G K L ...
```

Fig. 5. The ELK region, a highly conserved motif adjacent to maize homeodomains. Leucine or hydrophobic residues occur with periodicity in the ELK motif and are boxed. Amino acids between the periodic hydrophobic residues are predominantly charged or hydrophilic, suggesting the motif could form an amphipathic alpha helix. Invariant E, L, and K residues are shaded. Sequences for R7, P11, P15, and B151 are incomplete.

ELK region could, however, form a novel amphipathic alpha helix with an offset hydrophobic face (not shown). Secondary structure predictions assign the region alpha-helical character, and the high density of paired charged residues could help stabilize the helix. The *Arabidopsis Antp*-like homeobox genes *Athb-1* and *Athb-2* (Ruberti et al., 1991) encode a classic leucine zipper type motif that is also fused to the homeodomain. The adjacent leucine zipper is located on the C-terminal side, however, in contrast to the N-terminally located ELK motif.

Transcription factors are thought to be composed of modules that supply discrete or cooperative functions (Frankel and Kim, 1991; Ptashne, 1988). Accordingly, some HD proteins from animal systems also contain other structural motifs of known or unknown function (Bopp et al., 1986; Herr et al., 1988; Nicosia et al., 1990; Sive and Cheng, 1991). We suggest the ELK motif provides a function separable from that of the adjacent homeodomain. As a distinct motif, the ELK region could facilitate coiled-coil type protein–protein interactions, perhaps by acting as a dimerization interface between similar *Kn1*-like homeoproteins. On the other hand, the juxtaposed ELK region could act as a cooperative module that is functionally inseparable from the HD. Its absence, however, from nonplant, *Kn1*-like HDs (Fig. 4B) argues against a requirement for modular cooperativity, but does not exclude the possibility. In either case, the plant-specific occurrence of the ELK motif suggests that HD function in plants may be achieved by a novel mechanism. Elucidating the role of the well-conserved regions found adjacent to plant HDs should enrich our understanding of the functional modules within this important class of regulatory proteins.

CONCLUSION

The analysis of maize *Kn1*, in combination with research on other experimental systems, shows that homeobox genes are involved in developmental processes in each of the animal, fungal, and plant kingdoms. While regulation of cell identity involves homeodomain proteins in all three kingdoms, much remains to be learned about the molecular mechanisms involved, especially in plants. The enormous plasticity displayed by plant cells suggests that plants require a special mechanism to specify and then maintain cell identity. *Kn1* is a component of what may be one such mechanism, and related homeobox genes could comprise additional elements. Plant *Kn1*-like homeoboxes define a potentially large class of transcription factors that will undoubtedly be involved in a variety of cellular processes. Future studies should reveal the extent of the role these homeobox genes play in generating the unique features that distinguish plant development. Once the functions of multiple individual plant homeodomains are understood, a more general comparison of the way homeodomain proteins are utilized in the regulatory architectures of diverse organisms will be possible.

ACKNOWLEDGEMENTS

We thank members of the lab for sharing their results prior to publication, and Ron Wells for preparing the manuscript. This research was supported by NSF grant DMB8819325 and by the USDA.

REFERENCES

Allen JD, Lints T, Jenkins NA, Copeland NG, Strasser A, Harvey RP, Adams JM (1991): Novel murine homeo box gene on chromosome 1 expressed in specific hematopoietic lineages and during embryogenesis. Genes Dev 5:509–520.

Bopp D, Burri M, Baumgartner S, Frigerio G, Noll M (1986): Conservation of a large protein domain in the segmentation gene *paired* and in functionally related genes of *Drosophila*. Cell 47:1033–1040.

Bryan AA, Sass JE (1941): Heritable characters in maize. J Hered 32:343–346.

Bürglin TR, Ruvkun G, Coulson A, Hawkins NC, McGhee JD, Schaller D, Wittman C, Müller F, Waterston RH (1991): Nematode homeobox cluster. Nature 351:703.

Burr B, Burr FA, Thompson KH, Albertson MC, Stuber CW (1988): Gene mapping with recombinant inbreds in maize. Genetics 118:519–526.

Damante G, Di Lauro R (1991): Several regions of *Antennapedia* and thyroid transcription factor 1 homeodomains contribute to DNA binding specificity. Proc Natl Acad Sci USA 88:5388–5392.

Davidson EH (1990): How embryos work: A comparative view of diverse models of cell fate specification. Development 108:365–389.

Frankel AD, Kim PS (1991): Modular structure of transcription factors: Implications for gene regulation. Cell 65:717–719.

Freeling M, Hake S (1985): Developmental genetics of mutants that specify *Knotted* leaves in maize. Genetics 111:617–634.

Gehring WJ (1987): Homeo boxes in the study of development. Science 236:1245–1252.

Gelinas D, Postlethwait SN, Nelson OE (1969): Characterization of development in maize through the use of mutants. II. The abnormal growth conditioned by the *Knotted* mutant. Am J Bot 56:671–678.

Gibson G, Schier A, LeMotte P, Gehring WJ (1990): The specificities of sex combs reduced and *Antennapedia* are defined by a distinct portion of each protein that includes the homeodomain. Cell 62:1087–1103.

Goldberg RB (1988): Plants: Novel developmental processes. Science 240:1460–1467.

Graham A, Papalopulu N, Krumlauf R (1989): The murine and drosophila homeobox gene complexes have common features of organization and expression. Cell 57:367–378.

Hake S, Freeling M (1986): Analysis of genetic mosaics shows that epidermal cell divisions in *Knotted* mutant maize plants are induced by adjacent mesophyll cells. Nature 320:621–623.

Hake S, Vollbrecht E, Freeling M (1989): Cloning *Knotted*, the dominant morphological mutant in maize using *Ds2* as a transposon tag. EMBO J 8:15–22.

Hayashi S, Scott MP (1990): What determines the specificity of action of Drosophila homeodomain proteins? Cell 63:883–894.

Herr W, Sturm RA, Clerc RG, Corcoran LM, Baltimore D, Sharp PA, Ingraham HA, Rosenfeld MG, Finney M, Ruvkun G, Horvitz HR (1988): The POU domain: A large conserved region in the mammalian *pit*-1, *oct*-1, *oct*-2, and *Caenorhabditis elegans unc*-86 gene products. Genes Dev 2:1513–1516.

Herskowitz I (1989): A regulatory hierarchy for cell specialization in yeast. Nature 342:749–757.

Holland PWH, Hogan BLM (1988): Expression of homeobox genes during mouse development: A review. Genes Dev 2:773–782.

Kamps MP, Murre C, Sun X-H, Baltimore D (1990): A new homeobox gene contributes the DNA binding domain of the t(1;19) translocation protein in Pre-B ALL. Cell 60:547–555.

Kelly M, Burke J, Smith M, Klar A, Beach D (1988); Four mating-type genes control sexual differentiation in the fission yeast. EMBO J 7:1537–1547.

Kenyon C, Wang B (1991): A cluster of *Antennapedia*-class homeobox genes in a nonsegmented animal. Science 253:516–517.

Kissinger CR, Liu B, Martin-Blanco E, Kornberg TB, Pabo CO (1990): Crystal structure of an engrailed homeodomain-DNA complex at 2.8 Å resolution: A framework for understanding homeodomain-DNA interactions. Cell 63:579–590.

Kuziora MA, McGinnis W (1989): A homeodomain substitution changes the regulatory specificity of the *Deformed* protein in *Drosophila* embryos. Cell 59:563–571.

Landschulz WH, Johnson PF, McKnight SL (1988): The leucine zipper: A hypothetical structure common to a new class of DNA binding proteins. Science 240:1759–1764.

Langdale JA, Lane B, Freeling M, Nelson T (1989): Cell lineage analysis of maize bundle sheath and mesophyll cells. Dev Biol 133:128–139.

Lewis EB (1978): A gene complex controlling segmentation in *Drosophila*. Nature 276:565–570.

Mann RS, Hogness DS (1990): Functional dissection of the Ultrabithorax proteins in *D. melanogaster*. Cell 60:597–610.

McDaniel CN, Poethig RS (1988): Cell-lineage patterns in the shoot apical meristem of the germinating maize embryo. Planta 175:13–22.

McGinnis W, Levine MS, Hafen E, Kuroiwa A, Gehring WJ (1984): A conserved DNA sequence in homoeotic genes of the *Drosophila Antennapedia* and bithorax complexes. Nature 308:428–433.

Melton DA (1991): Pattern formation during animal development. Science 252:234–241.

Murre C, McCaw PS, Baltimore D (1989): A new DNA binding and dimerization motif in immunoglobulin enhancer binding, *daughterless*, *MyoD*, and *myc* proteins. Cell 56:777–783.

Nicosia A, Monaci P, Tomei L, De Francesco R, Nuzzo M, Stunnenberg H, Cortese R (1990): A myosin-like dimerization helix and an extra-large homeodomain are essential elements of the tripartite DNA binding structure of LFB1. Cell 61:1225–1236.

Nourse J, Mellentin JD, Galili N, Wilkinson J, Stanbridge E, Smith SD, Cleary ML (1990): Chromosomal trnslocation t(1;19) results in synthesis of a homeobox fusion mRNA that codes for a potential chimeric transcription factor. Cell 60:535–545.

Patel NH, Martin-Blanco E, Coleman KG, Poole SJ, Ellis MC, Kornberg TB, Goodman CS (1989): Expression of *engrailed* proteins in arthropods, annelids, and chordates. Cell 58:955–968.

Poethig RS (1987): Clonal analysis of cell lineage patterns in plant development. Am J Bot 74:581–594.

Ptashne M (1988): How eukaryotic transcriptional activators work. Nature 335:683–689.

Qian YQ, Billeter M, Otting G, Müller M, Gehring WJ, Wüthrich K (1989): The structure of the *Antennapedia* homeodomain determined by NMR spectroscopy in solution: Comparison with prokaryotic repressors. Cell 59:573–580.

Ruberti I, Sessa G, Lucchetti S, Morelli G (1991): A novel class of plant proteins containing a homeodomain with a closely linked leucine zipper motif. EMBO J 10:1787–1791.

Ruvkun G, Finney M (1991): Regulation of transcription and cell identity by POU domain proteins. Cell 64:475–478.

Schneuwly S, Kuroiwa A, Gehring WJ (1987): Molecular analysis of the dominant homeotic *Antennapedia* phenotype. EMBO J 6:201–206.

Scott MP, Tamkun JW, Hartzell III GW (1989): The structure and function of the homeodomain. Biochim Biophys Acta 989:25–48.

Scott MP, Weiner AJ (1984): Structural relationships among genes that control development: Sequence homology between the Antennapedia, Ultrabithorax, and fushi tarazu loci of Drosophila. Proc Natl Acad Sci USA 81:4115–4119.

Sharman BC (1942): Developmental anatomy of the shoot of *Zea mays* L. Ann Bot 6:245–281.

Sinha N, Hake S (1990): Mutant characters of *Knotted* maize leaves are determined in the innermost tissue layers. Dev Biol 141:203–210.

Sive HL, Cheng PF (1991): Retinoic acid perturbs the expression of *Xhox.lab* genes and alters mesodermal determination in *Xenopus laevis*. Genes Dev 5:1321–1332.

Smith L, Greene B, Veit B, Hake S (1992): A dominant mutation in the maize homeobox gene, *Knotted-1*, causes its ectopic expression in leaf cells with altered fates. Development (in press).

Sommer H, Beltran J-P, Huijser P, Pape H, Lönnig W-E, Saedler H, Scwarz-Sommer Z (1990): *Deficiens*, a homeotic gene involved in the control of flower morphogenesis in *Antirrhinum majus*: the protein shows homology to transcription factors. EMBO J 9:605–613.

Sussex IM (1989): Developmental programming of the shoot meristem. Cell 56:225–229.

Sylvester AW, Cande WZ, Freeling M (1990): Division and differentiation during normal and *liguleless-1* maize leaf development. Development 110:985–1000.

Veit B, Vollbrecht E, Mathern J, Hake S (1990): A tandem duplication causes the *Kn1-0* allele of *Knotted*, a dominant morphological mutant of maize. Genetics 125:623–631.

Vollbrecht E, Veit B, Sinha N, Hake S (1991): The developmental gene *Knotted-1* is a member of a maize homeobox gene family. Nature 350:241–243.

Walbot V (1983): Morphological and genomic variation in plants: *Zea mays* and its relatives. In: Goodwin BC, Holder N, Wylie CG (eds): "Development and Evolution". Cambridge, England: Cambridge University Press, pp 257–277.

Wingender E (1990): Transcription regulating proteins and their recognition sequences. Clinical Revs in Eukary Gene Exp 1:11–48.

Yanofsky MF, Ma H, Bowman JL, Drews GN, Feldmann KA, Meyerowitz EM (1990): The protein encoded by the *Arabidopsis* homeotic gene *agamous* resembles transcription factors. Nature 346:35–39.

Evolutionary Conservation of Developmental Mechanisms, pages 125–140

9. Evolutionary Conservation of Developmental Mechanisms: Comparisons of Annelids and Arthropods

David A. Weisblat, Cathy J. Wedeen, and Richard Kostriken
Department of Molecular and Cell Biology, University of California, Berkeley, California 94720

INTRODUCTION

Morphological changes accompanying the emergence of new species and the divergence of existing ones must arise through changes in the development of one species relative to another. Thus, it has been clear for at least 100 years that comparing the development of different species is one approach to understanding evolutionary relationships. One aspect of this comparative approach is to understand which embryological phenomena constitute primitive characteristics uniting a large group of species and which are derived characteristics that set apart smaller groups or individual species.

Despite the utility of the comparative approach in studying evolution and development, there are at least two difficulties that limit its implementation. One is that the experimental advantages and disadvantages of individual kinds of animals have largely dictated both the experimental techniques used to study them and the developmental questions addressed. This in turn limits one's ability to make direct comparisons. How, for example, is one to compare genetic studies of *Drosophila* segmentation, cell lineage analyses of *C. elegans*, biochemical studies of cell cycle in *Spisula*, and classical embryological studies of dorsal-ventral axis determination in *Xenopus*, in a manner that sheds light on the issues of evolutionary relationships of development? Another difficulty of the comparative approach is that it requires the study of all the species or groups to be compared, as opposed to focussing on the study of one particularly tractable species; there is no way around this requirement if one is to achieve a broad understanding of development. However, theoretical insights or experimental techniques achieved most readily from research on one organism have often stimulated and facilitated research on others and this approach might finally permit useful comparisons among organisms of diverse origins.

One area in the evolution of development that has been the object of considerable speculation is the origins of segmentation and the evolutionary relationship between segmented and unsegmented protostome phyla. Are arthropods, annelids, and mollusks monophyletic groups? And if so, what are their evolutionary relationships with each other and with the other protostome phyla? Did segmentation arise only once? If so, then what "primitive" (i.e. general) mechanisms underlie the radically different cellular processes leading to segmentation in, for example, insects and leeches? These questions are especially difficult to address from the fossil record alone because of the intrinsically poor preservation of soft-bodied animals. A good measure of our collective uncertainty in this area is the number of mutually exclusive phylogenetic "trees", each supported by well-reasoned and convincing arguments (e.g., see Willmer, 1990).

New information regarding phylogenetic relationships extending back hundreds of millions of years is starting to emerge from the comparison of nucleic acid sequences of very highly conserved genes, such as the ribosomal genes. Such data does not yield unique solutions; nonetheless, it seems likely that, as the sequence data base is expanded both in terms of the number of genes compared and the number of species examined within each taxon, a consensus will emerge regarding both the true composition of, and the approximate time since the last common ancestors of, the modern phyla.

The simple comparison of gene sequences cannot provide much information about either the body plan or the development of the ancestral species, however. For that, we must still rely on comparisons between currently existing species. Such comparisons remain quite subjective, especially when comparing highly diverged groups. But here, too, new molecular approaches are emerging, based on the discovery that certain developmental regulatory genes, identified initially by genetic techniques, most frequently in *Drosophila*, have close homologs in other species. This permits one to identify homologs of these genes in other species and analyze their roles in development.

Before doing so, however, it is essential to distinguish three different levels of homology: the most basic is *sequence homology*, i.e. similarities between two or more nucleic acid coding sequences; at an intermediate level is *biochemical homology*, i.e. the conservation of biophysical properties and therefore the most proximal biological function between two or more sequence homologs. For example, several homeodomain proteins have been shown to serve as transcription factors in *Drosophila* (Han et al., 1989). But what concerns us here is a third level of homology, in which the developmental function of two or more sequence homologs is also conserved. *A priori*, it seems that this would in most cases entail the phylogenetic conservation not just of single genes, but rather of sets of genes interacting with one another in reciprocal or hierarchical manner. For that reason, we define this highest level of homology as *syntagmal homology* (*syntagma*; Garcia-Bellido, 1985).

Here, after a brief introduction to the development of the family of leeches used in our research, we describe recent results from our studies of an *engrailed*-class gene in leech (*ht-en*) which suggest that this gene is a functional syntagmal homolog of the *Drosophila engrailed* gene. In doing so, we hope to illustrate the potential of a comparative approach in assisting our understanding of how evolution has produced the observed phyletic diversity through the progressive modification of regulatory processes.

AN OVERVIEW OF LEECH EMBRYOGENESIS

The details of the early cleavages and the general outlines of the overall development of glossiphoniid leeches like *Helobdella*, first described by C. O. Whitman (1878, 1887, 1892), have been amended and extended in recent years (Fig. 1). Schemes for blastomere nomenclature and staging embryos have been devised and amended (e.g. Fernandez, 1980; Bissen and Weisblat, 1989). The 0.5-mm eggs of *Helobdella triserialis* are fertilized internally, but do not begin cleaving until after they are laid in cocoons on the ventral aspect of the parent (stage 0). Embryos can be removed from the cocoons at any time and raised to maturity in simple salt solutions. Cleavages are stereotyped and many are unequal. Thus, many blastomeres can be individually identified by their size and position in the embryo, the order in which they arise, and by the segregation of domains of yolk-deficient cytoplasm.

During cleavage (stages 1–6), three distinct classes of blastomeres are formed, *macromeres, teloblasts*, and *micromeres*. The macromeres, A''', B''' and C''', are the three largest cells in the embryos by stage 6; they provide the foundation upon which the morphogenetic movements of embryogenesis take place, and are eventually enveloped by the gut and digested during stages 9–11. The teloblasts are five bilateral pairs of large stem cells derived from the D quadrant of the embryo (macromere D' of the 8-cell stage). Teloblasts, designated M, N, O/P, O/P and Q, ultimately give rise to all the segmentally iterated cells in the leech body, as will be described in greater detail below. The third class of cells is the micromeres, 25 small cells produced at various points during stages 4–6 by highly unequal cell divisions. At the end of stage 6, micromeres are found in a cluster called the *micromere cap* around the animal pole of the embryo.

Leeches are members in good standing of the phylum Annelida. Accordingly, their mesodermal and ectodermal tissues are organized into highly stereotyped, repeating subunits or *segments* throughout most of their length, and the developmental origins of the segments yield an interesting point of comparison between annelids and arthropods. In glossiphoniid leeches, segmentation occurs over an extended period of time during stages 7–10; segments arise and mature, as shown schematically in Figure 1, from what is in effect a posterior growth zone composed of the 10 teloblasts. One pair of teloblasts (M) generates segmental mesoderm, and four other pairs of teloblasts (N, O/P, O/P and Q) generate segmental ectoderm. During stages 7 and 8, each teloblast undergoes a series of

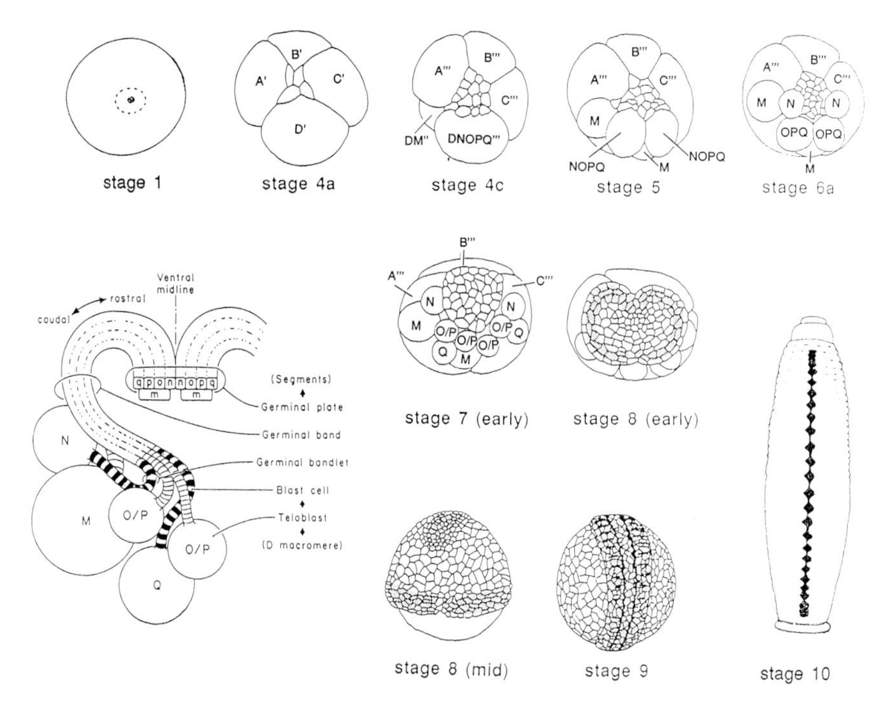

Fig. 1. Summary of leech development. Several stages of embryogenesis are illustrated, viewed from the animal pole or future dorsal aspect of the embryo, except for stages 9 and 10, which are ventral views. The egg is fertilized internally, but remains in meiotic arrest until it is laid (stage 1). After the two polar bodies have formed (small circles), cytoplasmic rearrangements generate domains of yolk-free cytoplasm, called teloplasm (dashed circle), at the animal and vegetal poles of the zygote. The third cleavage is highly unequal, yielding four macromeres and four micromeres (stage 4a); teloplasm is passed selectively to the D′ macromere. Macromeres A′, B′, and C′ each generate two more micromeres (stage 4c). Macromere D′ divides almost equally, to form precursors of segmental ectoderm (DNOPQ) and mesoderm (DM). During stages 5–6, these two cells undergo further stereotyped cleavages, producing five bilateral pairs of teloblasts, M, N, O/P, O/P, and Q, as well as additional micromeres. Each teloblast produces a coherent chain (bandlet) of segmental founder cells, called blast cells, by a series of highly unequal divisions. On each side the bandlets merge to form germinal bands (shown by stippling in drawings of stage 8–9), within which the bandlets occupy stereotyped positions relative to one another. During stage 8, the bands move across the surface of the embryo, gradually coalescing along the ventral midline from anterior to posterior to form the germinal plate, from which the definitive segmental tissues arise during stages 9–10. During stage 8, a micromere-produced epithelium (mosaic outlines) covers the bandlets and the dorsal surface of the embryo. During stages 9–10, the germinal plate expands dorsolaterally and gradually closes along the dorsal midline, displacing the provisional epithelium and forming the body tube of the leech. The ventral nerve cord, made up of segmentally iterated ganglia arises at the ventral midline of the germinal plate (filled contours in the stage 10 embryo) The yolky remnant of the macromeres and teloblasts is enveloped by the developing gut and digested. An enlargement of part of a stage 8 embryo is shown in the lower left of the figure, indicating the relative positions of the teloblasts and their blast cell progeny. Blast cells generated from N and Q lines are alternately colored black and white to denote that two blast cells are required to give rise to a single segmental complement of definitive progeny in these lineages.

several dozen unequal divisions at the rate of about one per hour (Wordeman, 1982), generating a column (*bandlet*) of *primary blast cells*. Ipsilateral bandlets come into parallel arrays called *germinal bands*, within which the bandlets occupy stereotyped positions; the bandlets are designated m, n, o, p, and q, as shown in Figure 1.

The left and right germinal bands are in contact with each other via their distal ends at the future head of the embryo and are separated from each other along the future dorsal side by a temporary epithelium derived from the cells of the micromere cap. This micromere-derived epithelium also covers the germ bands themselves. As more blast cells are budded off by the teloblasts and the germinal bands lengthen, they move across the surface of the embryo and gradually coalesce progressively from anterior to posterior along the future ventral midline into a structure called the *germinal plate* (stage 8). The micromere-derived epithelium expands concomitantly, continuing to cover the germ bands and the area behind them with a squamous epithelium. Beneath the micromere-derived epithelium in the area between the bandlets lie muscle fibers of mesodermal (M teloblast) origin (Weisblat et al, 1984). Together these two cell layers constitute the *provisional integument*, which serves as a temporary body wall of the embryo, pending the generation of a definitive body wall by the proliferation of cells in the germinal plate.

During stage 9, segments become apparent, first through the expansion of mesodermal hemisomites and later through the appearance of segmental ganglia, both tissues arising in a rostro-caudal progression from the *germinal plate*. During stage 10, the edges of the expanding germinal plate meet along the dorsal midline, closing the body tube of the leech, and during stage 11, segmental tissues differentiate to a state approximating their mature form.

Segmental tissues arise in a highly determinate manner by the stereotyped divisions of the blast cells within the germinal bands and germinal plate; older blast cells in each bandlet contribute to more anterior segments, and this accounts for the pronounced rostro-caudal temporal gradient seen throughout development (Braun and Stent, 1989). Each bandlet contributes a distinct subset of segmentally iterated neurons and nonneuronal progeny to the mature leech (Weisblat et al., 1984). Each m, o, and p blast cell generates a full *segmental complement* of definitive progeny for its cell line. But individual m, o, and p clones extend over more than one segment and thus interdigitate extensively with anterior and posterior clones of the same lineage (Weisblat and Shankland, 1985). Thus, in contrast to *Drosophila* (Crick and Lawrence, 1975; Garcia-Bellido, 1985), segments in leeches are not polyclones (the sum of several clones) derived from a small set of founder cells determined early in development. In the N (and Q) cell lines, *two* classes of blast cells, nf and ns (and qf and qs), arise in exact alternation, each contributing a specific subset of cells to a segmental complement of definitive N (or

Q) progeny (Weisblat et al., 1984; Zackson, 1984; Bissen and Weisblat, 1987). Within the N cell line in particular, the anterior portion of a segmental complement is generated almost entirely from nf-derived progeny, whereas the posterior portion is generated almost entirely from the nf-derived progeny (Bissen and Weisblat, 1987). The nf and ns clones within a segment intermingle slightly, but do not extend beyond the segment borders (Braun and Stent, 1989).

IDENTIFICATION OF AN *ENGRAILED* SEQUENCE HOMOLOG IN *HELOBDELLA TRISERIALIS*

A prominent feature of segmentation in *Drosophila* is the establishment of subdivisions called anterior and posterior *compartments* within the ectoderm of each prospective segment. The compartments are polyclones and are characterized by the property that cells in one compartment can mingle among themselves but have not been observed to mingle with cells in adjacent compartments (Garcia-Bellido et al., 1973; Garcia-Bellido, 1975). At about the same time that compartment boundaries are established, a circumferential band of cells in the embryo corresponding to the posterior compartment of each segment differentially expresses the gene *engrailed* (Kornberg, 1981b; Kornberg et al., 1985; DiNardo et al., 1985). The definition of anterior and posterior compartment fates is not simply a question of whether or not the cells express *engrailed*, however. A complex molecular network regulates the expression of most genes involved in *Drosophila* segmentation, including *engrailed*. But *engrailed* mutants show signs that the identity of the posterior compartment has been lost, including absence of restrictions to cell mingling and the apparent replacement of cuticular structures appropriate for the posterior compartment by those normally associated with the anterior compartment (Lawrence and Morata, 1976; Nusslein-Volhard and Weischaus, 1980; Kornberg, 1981a; Lawrence and Struhl, 1982). Later in development, *engrailed* is expressed in a subset of neurons within the segmental ganglia, where it presumably participates in the specification of a particular subset of neuronal phenotypes. Thus, as do several other of the segmentation genes that have been studied, *engrailed* appears to have two functions that are at least temporally distinct in *Drosophila* development, one during segmentation and another during neurogenesis.

The *engrailed* protein belongs to a family of transcription factors called homeodomain proteins. Members of this family are marked by a 60 amino acid DNA-binding domain called the homeodomain. Homeodomain proteins appear to occur in all higher eukaryotes, and within this large gene family, subfamilies can be recognized which have especially prominent sequence conservation within and outside the homeodomain. One such subfamily, the *en*-related genes, includes *engrailed* and *invected* genes in

Drosophila and their sequence homologs in other species. Thus, we can inquire as to the extent of the homology (as defined in the introduction) between *en*-related genes in various organisms, using a comparative approach.

A first step to answering this question is to examine the expression patterns of the *en*-related genes in the species of interest, and this process was facilitated by the discovery of a monoclonal antibody, mab4D9, that recognizes a stretch of roughly 10 amino acids within the homeodomain of *engrailed* and *invected* (Patel et al., 1989b). The monoclonal antibody, mab4D9, cross-reacts with many species, because this epitope is well conserved among *en*-related proteins (Fig. 2) (Patel et al., 1989b). In vertebrates, mab4D9 labels prospective anterior neural tissues, specifically in the region that will lie at the border between the midbrain and hindbrain (Patel et al., 1989b; Hemmati-Brivanlou and Harland, 1989; Gardner et al., 1988). Segmentally iterated expression is observed in muscle tissues, but only after somites are well defined (Patel et al., 1989b), by which point any gene that is not expressed ubiquitously will appear in an iterated pattern. Thus, it was suggested that the primitive or primordial function of *engrailed*-class genes was in neural differentiation and not in segmentation.

The same antibody, mab4D9, was also used to infer the pattern of *engrailed* expression in leech. In *Helobdella* embryos, mab4D9 identifies neurons in the subesophageal ganglion but does not identify segmentally iterated elements (Weisblat et al., 1989; Patel et al., 1989b). These results supported the notion that the original function of *engrailed* was in neurodifferentiation. It was also concluded that (1) annelids and arthropods were not particularly

```
                                1                  2                3
Ht-en      QDEKRPRTAFTGDQLARLKREFSENKYLTEQRRTCLAKELNLNESQIKIWFQNKRAKMKKASGVKNQLALQLMAQGLYNHSSSSSSSSSSSSSIFLLA
E30        PE---------SAE----------A--R----R--QQ-SRD-G-T-A--L----------I-----Q--P--------------TVPVDEDGEEI
E60        PE---------S-E----------A--R----R--QQ-SRD-G---A--L----------I-----Q--P--------------TVPLTKEEEEQ
D.m.-en    N----------SSE----------N--R----R--QQ-SS--G---A-------------I--ST-S--P-------------TTVPLTKEEEELEMRMNGQIP
D.m.-inv   P-D--------S-T-------H--N--R----K--QQ-SG--G---A-------------L--S--T--P--------------TIPLTREEEELQELQE-ASARAALEPC
S.U.-en    A----------SAS--Q---Q--QQSN--------RS-----T-S---------------I-----L--D--R-----------TVPLEAD-MDT
Z.F.-EN    KED---------A---Q---A--QT-R-------AQS--Q--G-----------------I--------G--IR----------TT-KEDK-D-D
Xl-En2     KED---------A---Q------QT-R--------QS--Q--S-----------------I---T-N--S---H----------TT-KDGK-DSE
Mo-En1     KED---------AE--Q---A--QA-R-I------QT--Q--S-----------------I---T-I--G---H----------TTTVQDKDE-E
Mo-En2     KED---------AE--Q---A--QT-R--------QS--Q--S-----------------I---T-N--T--VH----------TTAKEGK-D-E
Hu-EN2     KED---------AE--Q---A--QT-R--------QS--Q--S-----------------I---T-N--T--VH----------TTAKEGK-D-E
                                                ***********
                                                  mab4D9
```

Fig. 2. Comparison of conceptual translation products from some *en*-class genes. The 98 amino acids of the available *ht-en* sequence are designated by the one letter code, on the top printed line. The sequence begins with the one amino acid N-terminal to the homeodomain and ends with the last amino acid before the first termination codon. The homeodomain is underlined. Residues putatively participating in alpha helixes 1–3 (Kissinger et al, 1990) are designated by bold lines above the *ht-en* sequence. The sequence is aligned with sequences of *en*-class genes from honeybee (E30, E60), *Drosophila* (en, inv), sea urchin (S.U.-en), zebrafish (ZF-EN), frog (X1-En2), mouse (Mo-En1, Mo-En2), and human (Hu-EN2) (Dolecki and Humphries, 1988; Joyner and Martin, 1987; Fjose et al., 1988; Darnell et al., 1986; Walldorf et al., 1989; Poole et al., 1989; Logan et al., 1989; Hemmati-Brivanlou et al., 1990). Dashes represent amino acids identical to those of *ht-en*. The region of the 11 amino acid epitope recognized by mab4D9 in *Drosophila* (Patel et al., 1989) is designated by asterisks below the figure.

close phylogenetically and (2) the function of *engrailed* in segmentation arose after the separation of the arthodpod phylum (Patel et al., 1989b).

The staining of leech embryos by mab4D9 was relatively weak, however, and difficult to reproduce (D. Price personal communication). At about the same time, we cloned a fragment of a leech *engrailed* sequence homolog (*ht-en*) by low stringency hybridization from a *Helobdella* genomic library and sequenced it (Wedeen et al., 1991). From this it was learned that the sequence of the cloned fragment bears a nonconservative substitution in the region coding for the epitope that is recognized by the monoclonal antibody mab4D9 (Fig. 2).

Attempts to identify additional *en*-class genes in *Helobdella* by screening 10 genome equivalents of a leech genomic library at low stringency, with a *Drosophila en* homeobox probe (Wedeen et al., 1991), and by low stringency hybridization of a homeobox-containing fragment from *ht-en*, to Southern blots of *Helobdella* DNA (under conditions that revealed both *en*-class genes in *Drosophila*), were unsuccessful (Wedeen and Weisblat, 1991). Thus, although we cannot definitively rule out the existence of a second *en* class gene in *Helobdella* on the basis of these negative results, it appears that *ht-en* is the only *en*-class gene in *Helobdella*. Moreover, on immunoblots, mab4D9 fails to detect *ht-en* under conditions in which the *Drosophila en* epitope is strongly bound. Thus, it seems that the antigen detected by mab4D9 in leech embryos is not *ht-en*, but rather some other non–*en*-class antigen.

To generate antibodies that would recognize the *Helobdella engrailed* protein, four tandem repeats of the homeodomain and adjacent carboxyl-terminal sequences were cloned in frame with the *E. coli* β-galactosidase (βgal) gene (pUR 278) (Ruther and Muller-Hill, 1983). The iterated *ht-en* sequence was employed to bias the rabbit's immune response toward the *ht-en* portion of the fusion protein. The βgal fusion protein was induced and purified, and was then used to immunize a rabbit; the resultant polyclonal antibodies directed against *ht-en* (α*ht-en*) were affinity purified from the crude antiserum.

EXPRESSION OF *HT-EN* IN HELOBDELLA TRISERIALIS

The expression pattern of *ht-en* as revealed by α*ht-en* bears no resemblance to the pattern obtained with mab4D9. Expression is first apparent during late stage 7, and occurs as a dynamic series of segmentally iterated patterns. Double-label experiments, combining immunohistochemical staining with microinjected lineage tracers, show that the segmentally iterated patterns comprise specific cells or small subsets of cells in each of the five bandlets at different times in development (D. Lans and C.J.W., personal communication). For example, the first cells to express *ht-en* are those designated p.ap in the p bandlet, when the p blast cell clone contains about 5 cells.

By late stage 8, clones of cells derived from primary mesodermal (m) blast cells in the anterior portion of the germinal plate have formed hollow blocks of tissues corresponding to hemisomites (Zackson, 1982). Within the ectoderm, aggregations of cells are starting to become apparent, but ganglia are not yet evident (Fernandez, 1980; Torrence and Stuart, 1986). Within these anterior segments, *ht-en* immunoreactive nuclei appear as segmentally iterated transverse stripes. In each segment, the stripe extends across the central portion of the germinal plate, with a discontinuity at the ventral midline. By early stage 9, in accord with the rostrocaudal progression of development, the stripes are present in posterior segments. Because of the rostrocaudal gradient of developmental time in the leech embryo, the temporal progression of *ht-en* expression can be inferred by observing more anterior, developmentally more advanced segments (Fig. 3). Initially, the stripe consists of two to three nuclei on either side of the midline (Fig. 3b, c). Over the next few hours, however, additional nuclei become immunoreactive, so that the stripe eventually contains six to seven nuclei on each side (Fig. 3e). When segmental ganglia first become apparent, the boundaries between adjacent ganglia are aligned with the stripes (Fig. 3b, c). Then, as development proceeds, *ht-en* disappears in all but two nuclei on each side. These two cells move laterally as the germinal plate expands (Fig. 3d).

Using lineage tracer in combination with immunohistochemical detection of *ht-en* expression on these later stage embryos reveals that the stripe of *ht-en*-positive (black) nuclei is derived entirely from the nf blast cell, whose definitive progeny contribute to the posterior half of the segment. The two pairs of nuclei in the stripe that persist in their expression (Fig. 3d) become cells nz1 and nz2, two peripheral neurons that were already known to be derived from the nf blast cells (Braun and Stent, 1989).

As the stripes fade, new *ht-en* expression appears in a subset of ganglionic cells and extraganglionic cells derived from various classes of blast cells, in a stereotyped spatiotemporal pattern. Most of these cells maintain their expression at least as late as stage 11. Whether or not all the extraganglionic cells are neurons remains to be determined, as do the exact neuronal identities of the ganglionic neurons that express *ht-en*.

DISCUSSION

The pattern of *en*-class protein expression reported here for *Helobdella*, including both segmentally iterated stripes before ganglion formation and a segmentally iterated subset of presumptive neurons later in development, is strikingly reminiscent of the patterns of expression of *en*-class proteins previously described for arthropods (DiNardo et al., 1985; Brower, 1986;

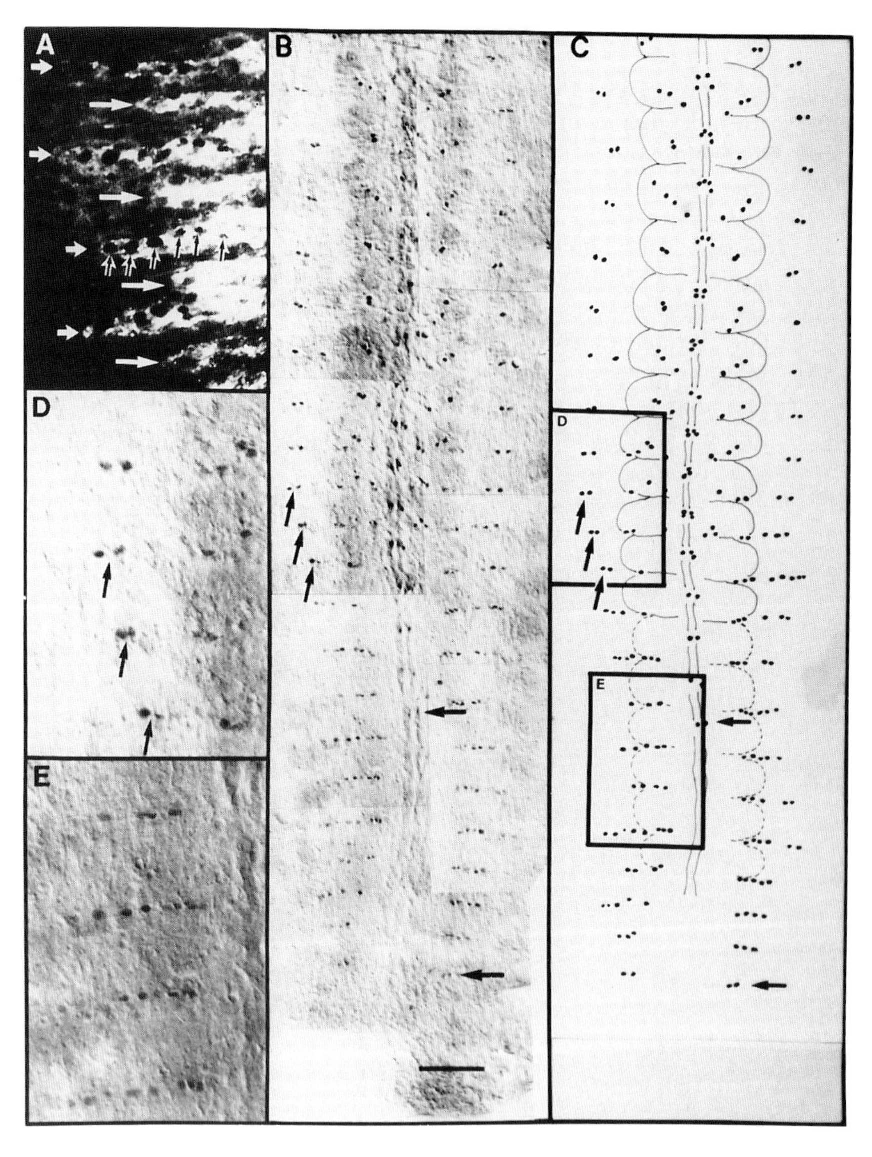

Fig. 3. *Ht-en* expression in the germinal plate. **A**: Confocal section of four half-segments from a stage 9 embryo in which an N teloblast was injected with RDA at early stage 7; anterior is up, the ventral midline is to the right. **B** and **C**: Bright-field photomontage and tracing of 22 segments in a dissected germinal plate. **D** and **E**: Higher magnification of fields shown by boxes in C. Anterior is up, the ventral midline is at the center. Developing ganglia are designated by the solid outline; areas where ganglia would soon develop are outlined by dashes. Lower right arrow designates the earliest *ht-en* expression in the stripe as two bilateral pairs of nuclei. Anterior to the lower arrow, stripes of nuclei mark the borders between future ganglia (see E). As expression arises first, in two pairs of nuclei along the ventral midline in newly developed ganglia (upper right arrow), expression begins to fade in the stripe, except for two cells (nzl and nz2; left arrows and D). More anteriorly, additional ganglionic nuclei express *ht-en*. Scale bar-26 μ in A, 50 μ in B, 23 μ in D and E.

Patel et al., 1989a; Fleig, 1990). We interpret these similarities as indicating that a pattern of dual expression of an *en*-class protein is a primitive trait shared by annelids and arthropods. There are two possible corollaries to this observation, depending on what other assumptions are made:

1. To the extent that one assumes that the utilization of *engrailed*-related genes in early development is evolutionarily recent, then the annelid and arthropod phyla must be correspondingly closely related, perhaps even sister taxa.
2. If one assumes that annelids are no more closely related to arthropods than to molluscs or other protostomes, then one can expect that the dual expression of *en*-class genes arose much earlier in evolution than previously supposed and that analogous dual expression of these genes will also be found in other protostome phyla. To us, this seems a more likely alternative, given the weight of molecular and classical taxonomic evidence (Field et al., 1988; Lake, 1989; Barnes et al., 1988).

These alternatives can be readily distinguished. For example, we predict that *engrailed*-related genes in onychophorans will also be expressed in a segmentally iterated pattern in early development, and our preliminary results indicate that this may be true (C.J.W. and P. Whitington, personal communication). The analysis of *en*-class gene expression in molluscs might be of greater interest, since most members of this phylum are not generally regarded as being segmented. Is the metamerism of the molluscan class polyplacophora (i.e. chitons) a secondarily derived trait, or are mollusks derived from a segmented ancestor common to annelids and arthropods? If the latter is true, we might observe dual expression patterns of *engrailed*-related genes in chitons reminiscent of those in *Helobdella* and *Drosophila*.

The observation of segmentally transverse rows of nuclei expressing *ht-en* in stage 8 to stage 9 *Helobdella* immediately brings to mind the stripes of cells expressing *engrailed* in the cellular blastoderm of *Drosophila* embryos. Moreover, the fact that all the cells in these rows are descended from just the nf (posterior) class of n blast cells suggests that the anterior/posterior compartments of *Drosphila* and the nf/ns blast cells in *Helobdella* may both derive from a segmental subdivision that was present in a common ancestor, consistent with the previously proposed notion of homology between the ns and nf blast cells in leech and the anterior and posterior compartments of arthropods (Bissen and Weisblat, 1987).

The segmentation processes of *Helobdella* and *Drosophila* differ extensively at the cellular level. The most obvious difference is the fact that *Helobdella* development proceeds via complete cleavages from the beginning, whereas in *Drosophila*, the first 13 rounds of nuclear proliferation occur in a syncytium. But there are many other significant differences as well. In *Helobdella*, for example, identifiable founder cells for hemisomites arise sequentially beginning with the 8th cell cycle roughly 15 h after egg deposition, as primary m blast cells; these immediately begin generating morphologically distinct clumps of cells (Zackson, 1982). In *Drosophila*, by contrast, morphologically distinct segmental structures arise roughly synchronously in the mesoderm about 3.5–4.5 h after fertilization, but this is during the 15th cell cycle (Campos-Ortega and Hartenstein, 1985). Thus, although embryogenesis proceeds at a much slower rate in *Helobdella* than in *Drosophila*, morphogenic aspects of segmentation in leech are markedly advanced in terms of cell generations.

Because the segmentation processes are so different in *Drosophila* and *Helobdella*, it has been very difficult to make meaningful comparisons of them. Now, however, given the remarkable similarities in the patterns of *en*-class gene expression, we have a starting point on which to base comparisons and speculations as to common mechanisms. In the early *Drosophila* embryo, the interface between the *en*-expressing cells and neighboring cells is necessary to establish the segmental and parasegmental borders (Martinez-Arias et al., 1988). It may be that the transient expression of *ht-en* in a segmentally iterated stripe is necessary to delineate the border between developing ganglia and/or other segmental structures. The *en* gene is eventually expressed in several rows of cells per segment, but there is no a priori reason why such an interface could not be achieved by a single row of cells, as observed in *Helobdella*. Resolution of this issue awaits empirical information as to the function *ht-en*. For this reason, the development of techniques for manipulating gene expression in leech embryos is of great interest.

Nonetheless, even without understanding the function of *ht-en* in *Helobdella*, the similarities between the expression patterns of *en*-class genes in annelid and arthropod suggest that *engrailed* and *ht-en* are syntagmal homologs. If this is true, it follows from the introduction that we should be able to expand the comparative approach by asking whether or not syntagmal homology exists between annelids and arthropods for genes whose products regulate or are regulated by *engrailed* class genes. For example, in *Drosophila*, the segment polarity gene *wingless* is normally expressed in the circumferential band of cells immediately anterior to the band of cells expressing *engrailed* (van den Heuvel et al., 1989). Genetic analysis has demonstrated that *wingless* function is required for the normal maintenance of *engrailed* expression (DiNardo et al., 1988; Martinez-Arias et al., 1988). Another segment polarity gene, *cubitus interruptus Dominant (ci-D)*, is initially expressed in all the cells of the blastoderm but later resolves into a striped pattern within the segmental ectoderm that is precisely complementary to *engrailed* expression pattern (Orenic et al., 1990). The expression of *ci-D* in the absence of the *engrailed* function fails to resolve into stripes suggesting that *engrailed* normally functions to repress *ci-D* (Eaton and Kornberg, 1990). Thus, *wingless* and *ci-D* interact indirectly or directly with *engrailed* as regulator and target gene, respectively.

The *ci-D* gene codes for a presumptive transcription factor of the zinc finger class (Orenic et al., 1990), for which three human sequence homologs have been identified (Ruppert et al., 1988). The gene *wingless* codes for a cell-associated, secreted protein (van den Heuvel et al., 1989) having sequence homology to the murine *wnt-1* gene (van Ooyen and Nusse, 1984); both are members of the newly recognized gene family, called *Wnt*, whose protein products appear to be important in modulating intercellular communication (Olson et al, 1991). The existence of multiple sequence homologs in both these gene families facilitated the design of degenerate oligonucleotides with which to seek annelid sequence homologs.

As of now, a sequence homolog to *ci-D* has been identified in the leech (Fig. 4) and it will be surprising if a *wingless* sequence homolog is not found, given the existence of *Wnt* class genes in arthropods and vertebrates. The expression patterns of these genes should provide strong hints as to whether or not their developmental interactions with *engrailed* are also conserved in annelids, and are therefore, (presumably) ancestral to both phyla. Moreover, since leech neurons are individually identifiable, both from segment to segment and from one species to another, it should be possible to identify exactly which neurons express *ht-en*, using *Hirudo medicinalis*, a species whose neurons are readily accessible for physiological characterization (C.J.W., K. French, and W. Kristan, unpublished observations). We can then ask whether or not there is any phylogenetic conservation between leech and insect with regard to the *en*-regulated genes that are responsible for generating the final phenotype of the neuron (e.g., those coding for receptors, channels, or enzymes for neurotransmitter synthesis). By this approach, we can further explore the notion (Garcia-Bellido, 1985) that not only individual genes but syntagmas have been conserved throughout the evolution of development. If so, the next challenge will be to define the extent both horizontally (in terms of interacting networks of genes) and vertically (in terms of hierarchically acting genes) of conserved syntagmas, with the ultimate goal of understanding the level(s) at which significant evolutionary changes have occurred.

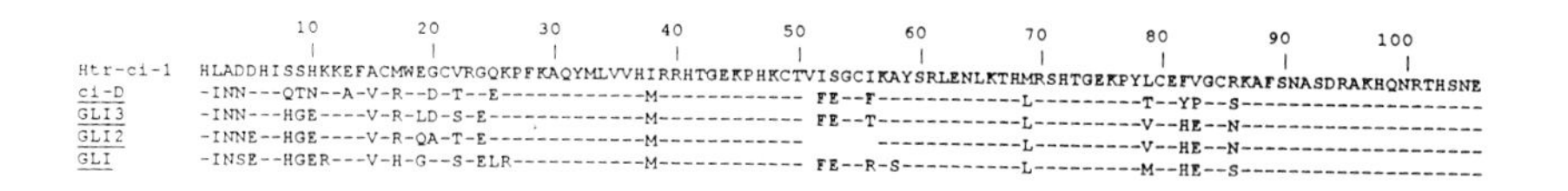

Fig. 4. Comparison of a portion of the conceptual translation product of *Htr-ci-1* with similar regions of *Drosophila cubitus interruptus-D* (*ci-D*; Orenic et al., 1990) and human *GLI* genes (Kinzler et al., 1988; Ruppert et al., 1988). Dashes indicate amino acids identical to *Htr-ci-1*.

ACKNOWLEDGMENTS

Parts of this work were supported by grants from the NIH, the NSF, and the March of Dimes Birth Defects Foundation. We thank Nipam Patel, David Jacobs and Deborah Lans for many helpful discussions.

REFERENCES

Barnes RSK, Calow P, Olive PJW (1988): "The Invertebrates: A New Synthesis." Oxford: Blackwell Scientific Publications.

Bissen ST, Weisblat DA (1987): Early differences between alternate n blast cells in leech embryos. Neurobiol 18:251–269.

Bissen ST, Weisblat DA (1989): The durations and compositions of cell cycles in embryos of the leech, *Helobdella triserialis*. Development 106:105–118.

Braun J, Stent GS (1989): Axon outgrowth along segmental nerves in the leech. I. Identification of candidate guidance cells. Dev Biol 132:471–485.

Brower DL (1986): *Engrailed* gene expression in imaginal discs. EMBO J 5:2649–2656.

Campos-Ortega JA, and Hartenstein V (1985): "The Embryonic Development of Drosophila melanogaster." New York: Springer-Verlag.

Crick FHG, Lawrence PA (1975): Compartments and polyclones in insect development. Science 189:340–347.

Darnell DK, Kornberg T, Ordahl CP (1986): J Cell Biol 103: 311a.

Davis CA, Noble-Topham SE, Rossant J, Joyner AL (1988): Expression of the homeobox containing gene En-2 delineates a specific region of the developing mouse brain. Genes Dev 2:361–371.

DiNardo S, Kuner JM, Theis J, O'Farrell PH (1985): Development of embryonic pattern in *D. melanogaster* as revealed by accumulation of the nuclear *engrailed* protein. Cell 43:59–69.

DiNardo S, Sher E, Heemskerk-Jongens J, Kassis JA, O'Farrell PH (1988): Two-tiered regulation of spatially patterned *engrailed* gene expression during *Drosophila* embryogenesis. Nature 332:604–609.

Dolecki GJ, Humphries T (1988): An *engrailed* class homeobox gene in sea urchins. Gene 64:21–31.

Eton S, Kornberg TB (1990): Repression of *ci-D* in posterior compartments of *Drosophila* by *engrailed*. Genes and Development 4:1068–1077.

Fernandez J (1980): Embryonic development of the glossiphoniid leech, *Theromyzon rude*: Characterization of the developmental stages. Dev Biol 76:245–262.

Field KG, Olsen GJ, Lane DJ, Giovanni SJ, Ghiselin MT, Raff EC, Pace NR, Raff RA (1988): Molecular phylogeny of the animal kingdom. Science 239:748–753.

Fjose A, Eiken HG, Njolstad PR, Molven A, Hordvik I (1988): A zebrafish *engrailed*-like homeobox sequence expressed during embryogenesis. FEBS Lett 231:355–360.

Fleig R (1990): *Engrailed* expression and body segmentation in the honeybee, *Apis melifera*. Roux's Arch Dev Biol 198:467–473.

Garcia-Bellido A, Ripoll P, Morata G (1973): Developmental compartmentalization of the wing disk of *Drosophila*. Nature New Biol 245:251–253.

Garcia-Bellido A (1975): In *Cell Patterning*, CIBA Foundation Symposium 29, eds. R. Porter and J. Rivers, Associated Sci Publishers, Amsterdam.

Garcia-Bellido A (1985): Cell lineages and genes. Phil Trans R Soc Lond B 312:101–128.

Gardner CA, Darnell DK, Poole SJ, Ordahl CP, Barald KF (1988): Expression of an *engrailed*-like gene during development of the early embryonic chick nervous system. J Neurosci Res *21*, 426–437.

Han K, Levine MS, Manley JL (1989): Synergistic activation and repression of transcription by *Drosophila* homeobox proteins. Cell 56:573–583.

Hemmati-Brivanlou A, de la Torre JR, Holt C, Harland RM (1991): Cephalic expression and molecular characterization of *Xenopus En-2*. Development 111:715–724.

Hemmati-Brivanlou A, Harland RM (1989): Expression of an *engrailed*-related protein is induced in the anterior neural ectoderm of early *Xenopus* embryos. Development 106:611–617.

Joyner A, Martin G (1987): *En-1* and *En-2*, two mouse genes with sequence homology to the *Drosophila engrailed* gene: Expression during embryogenesis. Genes Dev 1:29–38.

Kinzler KW, Ruppert JM, Bigner SH, Vogelstein B (1988): The GLI gene is a member of the Kruppel family of zinc finger proteins. Nature 332:371–374.

Kissinger CR, Liu B, Martin-Blanco E, Kornberg TB, Pabco CO (1990): Crystal structure of an *engrailed* homeodomain-DNA complex at 2.8 Å resolution: A framework for understanding homeodomain-DNA interactions. Cell 63:579–590.

Kornberg T (1981a): Compartments in the abdomen of *Drosophila* and the role of *engrailed* locus. Devl Biol 86:363–381.

Kornberg T (1981b): *engrailed*: A gene controlling compartment and segment formation in *Drosophila*. Proc Natl Acad Sci USA 78:1095–1099.

Kornberg T, Siden I, O'Farrell PH, Simon M (1985): The *engrailed* locus of *Drosophila*: localization of transcripts reveals compartment-specific expression. Cell 40:45–53.

Lake JA (1990): Origin of the Metazoa. Proc Natl Acad Sci 87:763–776

Lawrence PA, Morata G (1976): Compartments in the wing of *Drosophila*: A study of the *engrailed* gene. Devl Biol 50:321–337.

Lawrence P, Struhl G (1982): Further studies of the *engrailed* phenotype in *Drosophila*. EMBO J 1:827–833.

Logan C, Willard HF, Rommens JM, Joyner AL (1989): Chromosomal localization of the human homeo box-containing genes EN1 and EN2. Genomics 4:206–209.

Martinez-Arias A, Baker NE, Ingham PW (1988): Role of segment polarity genes in the definition and maintenance of cell states in the *Drosophila* embryo. Development 103:157–170.

Nusslein-Volhard C, Weischaus E (1980): Mutations affecting segment number and polarity in *Drosophila*. Nature 287:795–801.

Olson DJ, Christian JL, Moon RT (1991): Effect of Wnt-1 and related proteins on gap junctional communication in *Xenopus* embryos. Science 252:1173–1176.

Orenic TV, Slusarski DC, Kroll KL, Holmgren RA (1990): Cloning and characterization of the segment polarity gene *cubitus interruptus Dominant* of *Drosophila*. Genes Dev. 4:1053–1067.

Patel NH, Kornberg TB, Goodman CS (1989a): Expression of *engrailed* during segmentation in grasshopper and crayfish. Development 107: 201–212.

Patel NH, Martin-Blanco E, Coleman K, Poole SJ, Ellis MC, Kornberg TB, Goodman CS (1989b): Expression of *engrailed* proteins in arthropods, annelids, and chordates. Cell 58:955–968.

Poole SJ, Law ML, Kao F, Lau Y (1989): Isolation and chromosomal localization of the human En-2 gene. Genomics 4:225–231.

Ruppert JM, Kinzler KW, Wong AJ, Bigner SH, Kao F-T, Law ML, Seunez HN, O'Brien SJ, Vogelstein B (1988): The GLI-Kruppel family of human genes. Molec Cell Biol 8:3104–3113.

Ruther U, Muller-Hill B (1983): Easy identification of cDNA clones. EMBO J 2:761–770.

Torrence SA, Stuart DK (1986): Gangliogenesis in leech embryos: Migration of neural precursor cells. J Neurosci 6:2736–2746.

van den Heuvel M, Nusse R, Johnston P, Lawrence PA (1989): Distribution of the *wingless* gene product in *Drosophila* embryos: A protein involved in cell-cell communication. Cell 59:739–749.

van Ooyen A, Nusse R (1984): Structure and nucleotide sequence of the putative mammary oncogene *int*-1: proviral insertions leave the protein-encoding domain intact. Cell 39:233–240.

Walldorf U, Fleig R, Gehring WJ (1989): Comparison of homeobox-containing genes of the honeybee and *Drosophila*. Proc Natl Acad Sci USA 86:9971–9975.
Wedeen CJ, Price DJ, Weisblat DA (1991): Cloning and sequencing of a leech homolog to the *Drosophila engrailed* gene. FEBS Let 279:300–302.
Wedeen CJ, Weisblat DA (1991): Segmental expression of an *engrailed*-class gene during early development and neurogenesis in an annelid. Development 113 (in press).
Weisblat DA, Kim SY, Stent GS (1984): Embryonic origins of cells in the leech, *Helobdella triserialis*. Dev Biol 104:65–85.
Weisblat DA, Price DJ, Wedeen CJ (1989): Segmentation in leech development. Development 104 (Suppl.): 161–168.
Weisblat DA, Shankland M (1985): Cell lineage and segmentation in the leech. Phil Trans R Soc Lond B 312:39–56.
Whitman CO (1878): The embryology of *Clepsine*. Quart J Microscop Sci 18:215–315.
Whitman CO (1887): A contribution to the history of the germ layers in *Clepsine*. J Morphol 1:105–182.
Whitman CO (1892): "The Metamerism of *Clepsine*." Festschrift zum 70, Geburtstage R. Leukarts, pp 385–395.
Willmer P (1990): "Invertebrate Relationships: Patterns in Animal Evolution." Cambridge: Cambridge University Press.
Wordeman L (1982): Kinetics of primary blast cell production in the embryo of the leech, *Helobdella triserialis*. Honors thesis, Department of Molecular Biology, University of California, Berkeley.
Zackson S (1982) S (1982): Cell clones and segmentation in leech development. Cell 31:761–770.
Zackson S (1984): Segment formation in the ectoderm of a glossiphoniid leech embryo. Devl Biol 104:143–160.

Evolutionary Conservation of Developmental Mechanisms, pages 141–158

10. Inductive Signalling in *C. elegans*

Paul W. Sternberg, Russell J. Hill, and Helen M. Chamberlin
Howard Hughes Medical Institute, Division of Biology-California Institute of Technology, Pasadena, California 91125

INTRODUCTION

We now know that the invariance of *C. elegans* development (Sulston and Horvitz, 1977; Kimble and Hirsh, 1979; Sulston et al., 1983) arises in part from highly reproducible cell interactions (Sulston and White, 1980; Kimble, 1981; Sulston et al., 1983; Priess and Thomson, 1987; Thomas et al., 1990; Schnabel, 1991; Wood, 1991). Each cell is formed in an identical position within the developing organism and is therefore susceptible to the same set of intercellular signals. Some of these interactions involve signals among cells of equivalent developmental potential (lateral signalling) while other involve signals between cells of distinct developmental potential (induction). The invariant development and the small cell number of this nematode species allows study of cell interactions at the level of individual cells. Recent molecular evidence indicates that nematodes share many classes of proteins with vertebrates and insects, including proteins involved in signal transduction and transcriptional regulation (for example, Finney et al., 1988; Way and Chalfie, 1988; Burglin et al., 1989; Kamb et al., 1989; Freyd et al., 1990; Georgi et al., 1990; Gross et al., 1990; Hu and Rubin, 1990; Lochrie et al., 1991). Thus, from the point of view of developmental phenomenology and molecular mechanisms, nematodes provide a useful experimental system for the study of general properties of cell signalling.

Nematode postembryonic development has proven to be amenable to intensive developmental genetic analysis for several reasons. First, the postembryonic lineages can be followed by direct observation with a conventional light microscope (Sulston and Horvitz, 1977). This fact has allowed the effects of mutations to be studied at the level of single cells (Chalfie et al., 1981; Sulston and Horvitz, 1981). Second, mutations disrupting late developmental events have been readily isolated. Some mutations disrupt genes specific to postembryonic development. In many other cases, including most of those described in this article, relatively rare, tissue-specific alleles of essential genes have been isolated that preferentially affect certain aspects of development. Third, much of nematode postembryonic development comprises sexual maturation (Sulston and Horvitz, 1977; Kimble and Hirsh,

1979; Sulston et al., 1980), and hence reproductive structures. The internal self-fertilization of *C. elegans* hermaphrodites allows mutant hermaphrodites or males incapable of copulation to be propagated and analyzed (Hodgkin et al., 1979; Horvitz and Sulston, 1980). This property allows the genetic analysis of hermaphrodite vulval and male spicule development, the foci of this article.

The cells of interest for this discussion are members of small sets of oligopotent cells whose fates are specified in part by intercellular signals. One such set of cells is the six vulval precursor cell (VPC) "equivalence group." Three of the six VPCs normally generate the vulva, which the hermaphrodite uses for egg-laying and copulation (Fig. 1 and 2) (Sulston and Horvitz, 1977). The precursor cells that consititute two other equivalence groups, the Bα/Bβ and Bγ/Bδ pairs, generate many of the cells of the male spicules (Sulston et al., 1980), which are inserted into the vulva during copulation.

Genes required for inductive signalling have been identified by mutations that lead to a failure in vulval induction and hence a Vulvaless phenotype (Fig. 3) (Horvitz and Sulston, 1980; Sulston and Horvitz, 1981; Ferguson and Horvitz, 1985; Sternberg and Horvitz, 1989; Beitel et al., 1990; Han et al., 1990; Kim and Horvitz, 1990; Aroian and Sternberg, 1991). Since *C. elegans* hermaphrodites are internally self-fertilizing, the vulva is not essential for reproduction (Fig. 4). At a cellular level, many of the Vulvaless mutations result in too few VPCs generating vulval cells: All six VPCs have 3° fates and generate nonspecialized hypodermis (Fig. 3). In contrast, four or more VPCs generate vulval cells in Multivulva mutants. In many Multivulva mutants, such excessive vulval differentiation can occur even in the absence of the

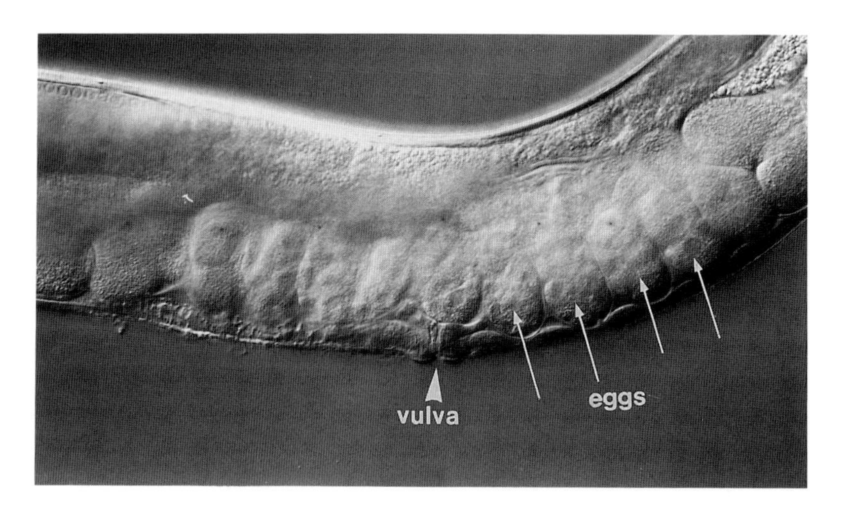

Fig. 1. Adult hermaphrodite vulva. Nomarski photomicrograph of a lateral view of the middle third of an adult hermaphrodite. Arrowhead points to opening of the vulva. Arrows point to four of the cleaving eggs present in the uterus.

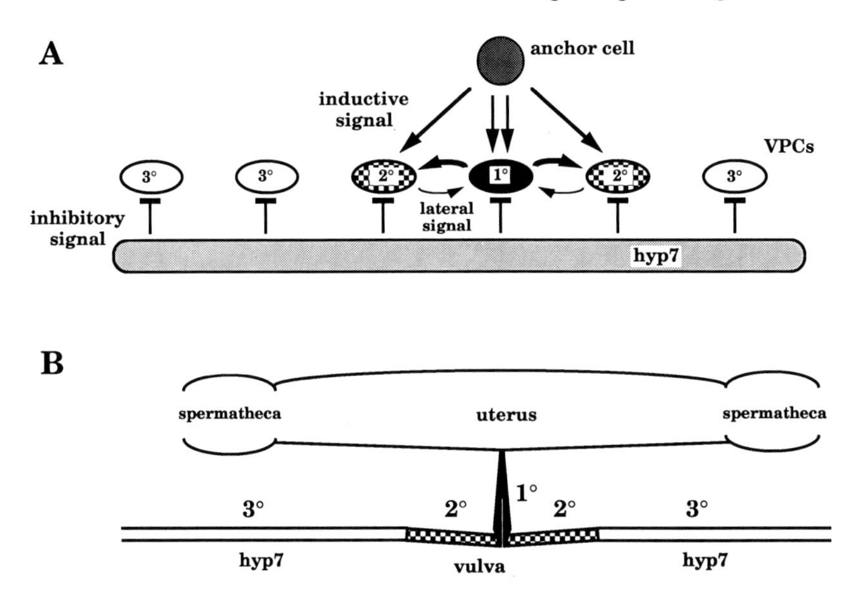

Fig. 2. Schematic of vulval induction. A: Three-signal model for vulval induction. Model based on proposals of Sulston and Horvitz (1977), Sulston and White (1980), Kimble, (1981), Sternberg and Horvitz (1986), Sternberg (1988), Herman and Hedgecock (1990), Horvitz and Sternberg (1991). B: Territories of adult vulva contributed by the three types of VPCs (Sulston and Horvitz, 1977). A single uterus is flanked by spermatheca and ovaries (not shown; see Hirsh et al., 1976).

gonadal anchor cell, which induces vulval formation (see below). The Vulvaless and Multivulva mutants have allowed the analysis of signalling pathways involved in vulval development.

The molecular cloning of genes based on their genetic map positions has been greatly facilitated in the past few years by the *C. elegans* genomic map (Coulson et al., 1986; Coulson et al., 1988) and by the ability to readily make transgenic nematodes by the microinjection of DNA into the syncytial gonad of *C. elegans* hermaphrodites (Stinchcomb et al., 1985; Fire, 1986). Most transgenes generated in this manner are maintained as semi-stable extrachromosomal arrays. These arrays, which are concatenates of the input DNA (Mello et al., 1991), are passed through meiosis and thus can be manipulated as genetic elements. In addition, as we shall discuss, the transgenes sometimes confer dominant phenotypes, and thus can be used in a test of epistasis to order the action of genes in a pathway.

THREE-SIGNAL MODEL FOR VULVAL DEVELOPMENT

The six VPCs are tripotent, each being able to express the 1°, 2°, or 3° fates (reviewed by Horvitz and Sternberg, 1991). The 1° and two 2° VPCs generate distinct sets of progeny cells that together form the vulva. The three 3° VPCs

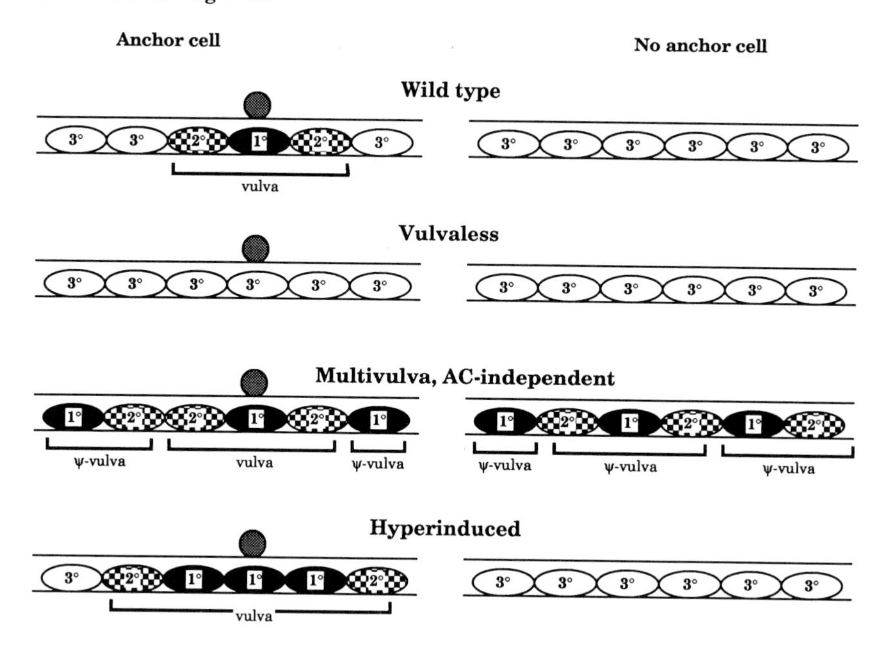

Fig. 3. Mutant types. Schematic of some key types of vulval induction mutants described in the text. The depicted patterns of cell types in the Multivulva and Hyperinduced are exemplary; there is some variability in the precise pattern (Sternberg, 1988; Sternberg and Horvitz, 1989; Aroian and Sternberg, 1991). Brackets indicate the cells that form the vulva or vulvalike protrusions (pseudovulva or ψ-vulva). AC=anchor cell.

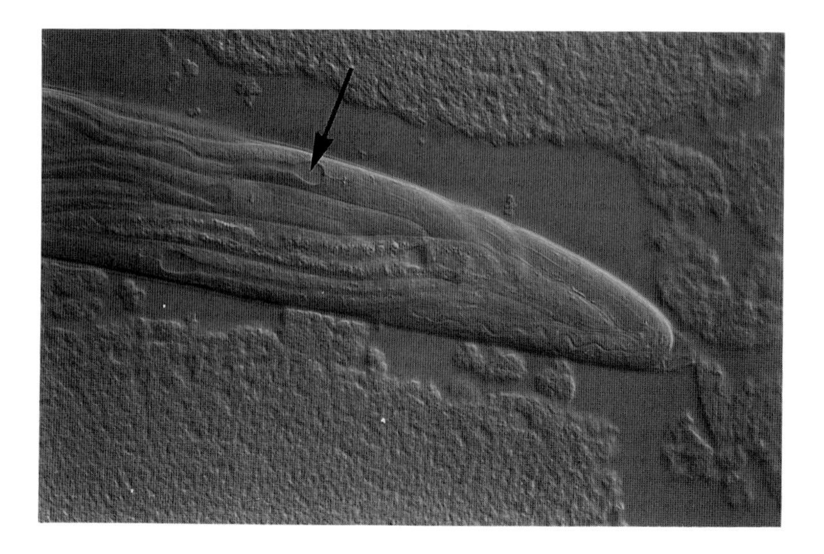

Fig. 4. Vulvaless hermaphrodite. Nomarski photomicrograph of a *let-23(sy1)* adult. Arrow points to the lumen of the intestine of an L1 larvae inside the cuticle of its mother.

each generate two progeny that fuse with the large hypodermal syncytium, hyp7. According to the simplest model, vulval development involves three intercellular signalling pathways whose combined action establishes the precise pattern of VPC fates, 3°-3°-2°-1°-2°-3° (Fig. 2); an inductive signal, an inhibitory signal, and a lateral signal.

An **inductive signal** from the gonadal anchor cell (AC) (Fig. 4) stimulates three VPCs to have the 1° or 2° fates (Fig. 2). These three VPCs proliferate and generate vulval cells. The induced VPC closest to the anchor cell becomes 1°, while the two adjacent VPCs become 2°. Destruction of the anchor cell eliminates the inductive signal (Fig. 3). If the anchor cell is absent, all six VPCs adopt the 3° fate and hence differentiate as nonspecialized hypodermis (Kimble, 1981). Many of the Vulvaless mutations disrupt inductive signalling and thus the genes they define may be involved in the generation of the inductive signal or the response to the inductive signal by the VPCs. The signal appears to be graded (Sternberg and Horvitz, 1986) and can act at a distance (Thomas et al., 1990).

An **inhibitory signal** from the hyp7 hypodermis prevents VPCs from proliferating and generating vulval cells in the absence of the inductive signal (Fig. 2 and 3) (Herman and Hedgecock, 1990). A mutation in *lin-15* results in a Multivulva phenotype such that all six VPCs generate vulval cells even if the anchor cell is absent (Ferguson and Horvitz, 1985; Ferguson et al., 1987; Sternberg, 1988; Ferguson and Horvitz, 1989; Sternberg and Horvitz, 1989). Mosaic analysis of the the *lin-15* locus suggests that it is required, not in the VPCs or in the anchor cell, but in other cells, most likely the hyp7 syncytial hypodermis (Herman and Hedgecock, 1990). The simplest interpretation of these data is that there is an inhibitory signal from hyp7 that acts on the VPCs. *lin-15* is one of a number of genes that act in this inhibitory signalling pathway (Ferguson and Horvitz, 1989).

In a wild-type hermaphrodite, the inductive signal from the anchor cell overrides the *lin-15* inhibitory signal to induce the three proximal VPCs to generate vulval cells. These two signals could act either in series or in parallel. One possibility is that the inductive signal could inhibit the action of *lin-15* (in series). Or, the inductive signal and *lin-15* could act antagonistically on the VPCs (in parallel). If the anchor cell inductive signal works by inhibiting the *lin-15* inhibitory signal, the presence or absence of the anchor cell should have no effect in a *lin-15* mutant. Two different observations indicate that the anchor cell affects the fates of VPCs in a *lin-15* mutant. First, the anchor cell influences the pattern of VPC fates in a *lin-15* mutant (Sternberg, 1988). In particular, P6.p, the VPC closest to the anchor cell, is always 1° in a *lin-15* mutant, but can be 2° in a *lin-15* animal whose anchor cell has been ablated. In other words, in a *lin-15* mutant, the anchor cell induces P6.p to become 1°. Therefore, the inductive signal can influence VPC fates in the absence of *lin-15* function. Second, some *let-60* mutations can ameliorate the *lin-15* defect such that vulval development is restored to gonad-dependence (Han

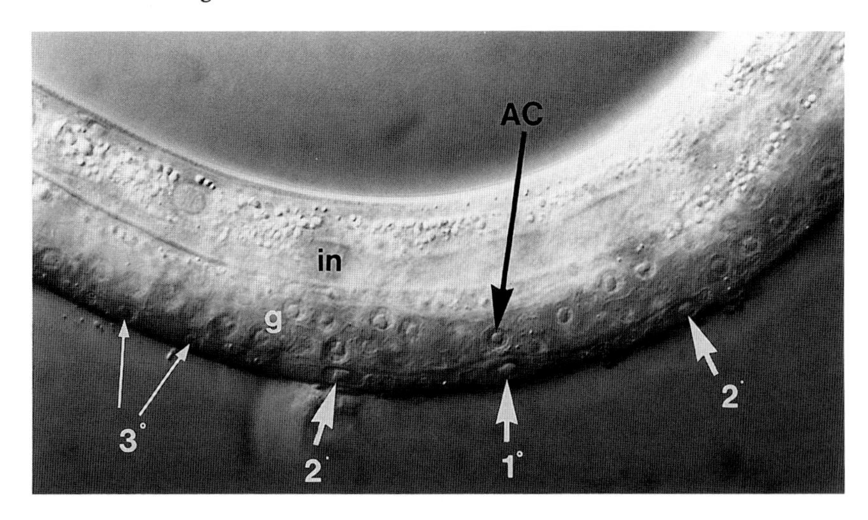

Fig. 5. Anchor cell and VPCs just after induction. The anchor cell nucleus (AC), indicated by the black arrow, is visible in the center of developing gonad (g) of this L3 stage larval hermaphrodite. The 1° and two 2° VPCs are morphologically indistinguishable at this stage; the large white arrows point to the nuclei of these VPCs. The 3° VPCs have just divided, and the daughters of one 3° VPC are indicated by the small white arrows.

et al., 1990). We therefore believe that the *lin-15* inhibitory signal and inductive signal act in parallel.

A **lateral signal** among the induced VPCs prevents adjacent cells from becoming 1° (Sternberg, 1988) (Fig. 2). A presumptive 1° VPC promotes the differentiation of is neighbors as 2° VPCs. The lateral signal could be one component of a feedback mechanism that would amplify small differences in the level of inductive signal received by the presumptive 1° and 2° VPCs to a larger differential in their response to inductive signal. The VPC receiving the most intense signal (or receiving the signal first) would become 1°. Such a mechanism would contribute to the precise 2°-1°-2° pattern of VPC fates. The *lin-12* gene, studied by Greenwald and her colleagues (Yochem et al., 1988; Seydoux and Greenwald, 1989; Greenwald and Seydoux, 1990) is a candidate for the receptor for lateral signal (Sternberg and Horvitz, 1989).

The three intercellular signalling pathways must be coupled in some fashion. First, as discussed above, the inhibitory signalling pathway acts to prevent action of the inductive signalling pathway in the absence of the inductive signal. Specifically, disruption by mutation of the inductive signal response pathway renders the inhibitory signalling pathway irrelevant (Han et al., 1990; Aroian and Sternberg, 1990). Moreover, in some circumstances, the inductive and inhibitory pathways act antagonistically in a quantitative manner: mutations that partially disrupt these pathways mutually suppress each other (Ferguson et al., 1987; Sternberg and Horvitz, 1989; Han et al., 1990). Therefore, the inductive and inhibitory signals may be integrated at the level of *let-23* in each VPC. Second, the lateral signalling pathway may be under control of the inductive signalling pathway (Sternberg and Horvitz, 1989). For example, *lin-12* mutations that appear to make lateral signalling constitutive cause VPCs to become 2° in the absence of inductive signal.

We focus on the inductive signalling pathway for the remainder of this discussion. We describe molecular genetic analysis of three genes that are necessary for vulval induction: *let-60* (*let*hal complementation group 60), *let-23,* and *lin-3* (cell *lin*eage defective gene 3). All three genes encode proteins that are similar to mammalian proteins involved in intercellular signalling or signal transduction. Genetic epistasis experiments indicate that these three genes act in a common pathway. *lin-3* acts through *let-23,* which acts through *let-60* to control vulval differentiation.

ROLE OF *let-60* RAS

Genetic analysis of the *let-60* locus led to the proposal that this gene acts as a component of a "switch" during vulval induction (Beitel et al., 1990; Han et al., 1990) (Fig. 6). Specifically, mutations that decrease *let-60* activity prevent VPCs from generating vulval cells, even in the presence of an inductive signal. Thus, *let-60* activity is needed for vulval induction. Mutations that increase *let-60* activity result in a Multivulva phenotype even if the anchor cell is destroyed by laser microbeam irradiation. In addition, extra copies of a wild-type *let-60* gene which are carried on an extrachromosomal array in transgenic hermaphrodites also result in a Multivulva phenotype (Han and Sternberg, 1990). Thus, increased *let-60* activity activates the vulval inductive pathway. Together, these observations indicate that the state of *let-60* activity specifies whether VPCs will generate vulval cells (active) or

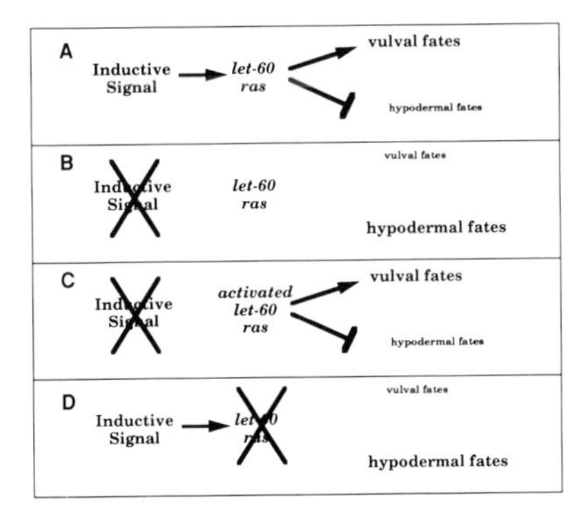

Fig. 6. *let-60* as a switch. A: In the wild type, indutive signal leads to the activation of *let-60* ras. B: In the absence of an inductive signal, *let-60* remains inactive. C: A mutation that activates *let-60* results in vulval differentiation in the absence of an inductive signal. D: A mutation that inactivates *let-60* prevents vulval induction.

nonvulval cells (inactive), and suggest that *let-60* is set in an active state by the action of the inductive signal. In addition to its role in vulval development, *let-60* is required for postembryonic growth and development: Apparent null mutations of *let-60* result in lethality during the L1 stage (Clark et al., 1988; M. Han and P.W. Sternberg, manuscript in preparation).

The inferred product of *let-60* is a Ras protein (Han and Sternberg, 1990). Of the first 164 amino acids, 136 are identical in human H-ras and *let-60* ras. Such conservation strongly indicates conserved biochemical function. Moreover, mutations altering particular residues of *let-60 ras* have analogous effects, as does cognate mutation in mammalian or yeast proteins (Beitel et al., 1990; M. Han and P.W. Sternberg, manuscript in preparation). Ras proteins bind guanine nucleotides and are thought to play a role in signal transduction in both yeast and mammalian cells. The crystal structure of the human H-Ras protein has been determined in both the GTP-and GDP-bound forms (reviewed by Barbacid, 1987). These structural studies indicate that Ras acts as a molecular switch, with two conformations that correspond to active (GTP-bound) and inactive (GDP-bound or empty site) forms. Like other GTP-binding proteins, Ras proteins undergo a cycle of guanine nucleotide exchange and hydrolysis (reviewed by Bourne et al., 1990; Bourne et al., 1991).

let-23 MAY BE THE RECEPTOR FOR THE INDUCTIVE SIGNAL

Genetic and molecular studies of the *let-23* gene suggest that it acts as a receptor for the inductive signal from the anchor cell. Genetic analysis of various classes of *let-23* mutations indicate that this gene is essential for both larval growth and vulva induction. *let-23* was originally identified by recessive larval lethal mutations (Herman, 1978; Sigurdson et al., 1984), and a partially lethal Vulvaless allele (Ferguson and Horvitz, 1985; Ferguson et al., 1987). Further analysis of *let-23* mutations indicated that complete elimination of *let-23* activity results in lethality during the L1 stage. This conclusion is based in part on two observations (Aroian and Sternberg, 1991). First, recessive, lethal alleles are the most common class of *let-23* mutations isolated in an F1 "non complementation" screen that could recover deletions of the locus. Second, recessive lethal alleles and a deletion of the locus result in similar phenotypes in *trans*-heterozygotes among various *let-23* alleles.

let-23 was cloned by correlation of the genetic and physical maps in the *let-23* region followed by DNA-mediated transformation to identify the *let-23* locus (Aroian et al., 1990). In contrast to *let-60*, extra copies of *let-23* do not result in a Multivulva phenotype. *let-23* encodes a homolog of the mammalian epidermal growth factor receptor (EGF-R) (Aroian et al., 1990). The overall architecture and specific sequence motifs of the inferred product of *let-23* are similar to the EGF receptor subfamily of receptor tyrosine kinases. This subfamily includes three human proteins and one *Drosophila* gene, as well as other vertebrate homologs. This subfamily is characterized by an N-terminal signal peptide, a large extracellular domain with two cysteine-rich regions, a single hydrophobic transmembrane domain, a tyrosine kinase domain, and a carboxy-terminal tail. The spacing of the cysteine residues in each cysteine-rich region is highly conserved among the mammalian, fish, insect, and nematode proteins. The kinase domains have approximately 40% identity and do not contain a "kinase insert," a characteristic of several other classes of receptor tyrosine kinases, such as the PDGF receptor. This high degree of conservation suggests that the *let-23* protein acts like a typical receptor tyrosine kinase. Thus, *let-23* protein presumably binds a ligand via its extracellular domain. Binding of ligand presumably results in transduction of a signal across the plasma membrane and activation of tyrosine kinase activity. An activated kinase would then phosphorylate target proteins to regulate their activity (reviewed by Ullrich and Schlessinger, 1990).

Rare *let-23* alleles and certain *let-23* heteroallelic combinations cause more VPCs to have vulval fates than in the wild type (Aroian and Sternberg, 1991). The pattern of extra vulval fates in *let-23(n1045)* animals is different from the Multivulva mutants. The lateral signalling process, proposed to normally prevent adjacent VPCs from becoming 1° (Sternberg, 1988), fails in such animals so that neighboring VPCs can be 1° (Aroian and Sternberg, 1991). The expression of the extra vulval VPC fates in *let-23(n1045)* animals depends on the inductive signal, unlike those in Multivulva mutants (Fig. 3). This phenotype is therefore referred to as "Hyperinduced."

Because *let-23* is required for vulval induction, but also can be mutated to cause excessive vulval induction, we believe *let-23* is intimately involved in control of the inductive signalling pathway. Based on its molecular architecture, *let-23* is likely to act as a receptor for an inductive signal from the anchor cell during vulval development. We will test this hypothesis by examining the cells in which *let-23* is expressed and in which it functions, for example, by mosaic analysis (Herman, 1989).

The order of *let-23* and *let-60* action has been suggested by epistasis experiments. Specifically, a *let-60* gain-of-function (*gf*) mutation was obtained as an extragenic suppressor of *let-23* Vulvaless phenotype (G. Jongeward, personal communication). Also, high-copy wild-type *let-60* in transgenic animals rescues a *let-23* Vulvaless defect (Fig. 7) (Han and Sternberg, 1990).

lin-3 MAY BE THE INDUCTIVE SIGNAL

Recent studies indicate that *lin-3* may encode the inductive signal made by the anchor cell. *lin-3* is required for vulval induction (Horvitz and Sulston, 1980; Sulston and Horvitz, 1981; Ferguson and Horvitz, 1985) and is necessary for the proper development of a set of tissues which is similar to the set of tissues that require *let-23*. Like *let-23* and *let-60, lin-3* is necessary for larval viability

A

chromosomal genotype	wildtype gene in transgene: none	⇑lin-3	⇑let-23	⇑let-60
+	WT	Muv	WT	Muv
lin-3(lf)	Vul	Muv	nd	Muv
let-23(lf)	Vul	Vul	WT	Muv
let-60(dn)	Vul	nd	nd	Muv

B

lin-3 ⟶ *let-23* ⟶ *let-60* ⟶ vulval fates

Fig. 7. Ordering the action of *lin-3, let-23,* and *let-60* using epistasis with dominant transgenes. A: Summary of transgenic data of Aroian et al., (1990), Han and Sternberg (1990), R.J. Hill and P.W. Sternberg (manuscript in preparation). Decreased function of *lin-3, let-23,* or *let-60* results in a Vulvaless phenotype. High copy of lin-3 or *let-60* genes in transgenic animals results in a Multivulva phenotype. If a multicopy array of *let-60* is crossed into a *lin-3* or *let-23* mutant, that strain will be Multivulva. If, however, a multicopy array of lin-3 is crossed into a let-23 mutant, that strain is Vulvaless. B: Pathway inferred from the epistasis data summarized in A. Vul=Vulvaless; Muv=Multivulva; WT=wildtype; nd=not determined; lf=loss-of-function; dn=dominant negative. See references for details.

(Ferguson and Horvitz, 1985) and for the proper development of the male spicules, as described below (H.M.C. and P.W.S. manuscript in preparation).

The *lin-3* locus was physically identified by transposon tagging, and DNA transformation was used to confirm and delimit the location of the *lin-3* gene (R.J.H. and P.W.S. manuscript in preparation). Multicopy *lin-3* transgenes, like *let-60* transgenes but unlike *let-23* transgenes, confer a dominant mulitivulva phenotype (Fig. 7). The level of vulval induction appears to be sensitive to the dose of *lin-3*: increased *lin- 3* copy number raises the level of vulval induction (R.J.H and P.W.S. manuscript in preparation), while mutations that decrease *lin-3* activity decrease the level of vulval induction (Sulston and Horvitz, 1981; R.J.H. and P.W.S., unpublished observations). The ability of the *lin-3* transgenes to stimulate vulval induction is blocked by mutations in *let-23*. Therefore, *lin-3* acts through *let-23* to promote vulval development. If *let-23* is the receptor for the inductive signal, *lin-3* might control production of a ligand for *let-23*, act as a ligand, or be a cofactor of *let-23* action. However, the action of *lin-3* and *let-23* can be separated into distinct steps by genetic epistasis experiments between a *lin-15* Multivulva mutation and *lin-3* or *let-23* Vulvaless mutations (Ferguson et al., 1987; R.J.H. and P.W.S., unpublished observations). Specifically, a *lin-15* mutation is epistatic to a *lin-3* but not to a *let-23* mutation, indicating that *lin-3* does not simply act with *let-23*. Therefore, *lin-3* is unlikely to be a cofactor of *let-23*.

The structure of *lin-3* suggests that is a ligand for *let-23*. The DNA sequence of the *lin-3* locus indicates that it encodes a growth factor precursor of the class that includes known ligands of the mammalian EGF receptor (EGF-R) (R.J.H. and P.W.S., manuscript in preparation). Ligands of the EGF-receptor include epidermal growth factor (EGF), transforming growth factor alpha (TGF-α), and vaccinia virus growth factor (VVGF). *lin-3* resembles these molecules in that they all possess a transmembrane domain and one or more EGF repeats (Fig. 8). *lin-3*, however, shows no sequence similarity to any of these growth factor precursors outside of the EGF repeat, or to any other described molecule. Thus, *lin-3* is possibly a novel member of the family of growth factors that contain EGF repeats. The single EGF repeat of pro-TGF-α and the membrane-proximal EGF repeat of pro-EGF are proteolytically processed to yield diffusible factors of 50 and 53 amino acids respectively (reviewed by Carpenter and Wahl, 1990). In the case of TGF-α it has been shown that the membrane bound form is also capable of activating the EGF receptor (reviewed by Massagué, 1990). On the basis of its genetics and predicted sequence, it is proposed that *lin-3* is the inductive signal made by the anchor cell to specify vulval fates.

We believe that the anchor cell inductive signal is diffusible. When all but one of the VPCs have been destroyed by laser microsurgery, the single remaining VPC can respond to the signal at a distance from the AC (Sternberg and Horvitz, 1986). In addition, vulval induction occurs in *dig-1* mutant animals in which the AC is displaced to the dorsal side of the animal away from the VPCs (Thomas et al., 1990). It is conceivable that the anchor cell inductive signal acts by direct cell contact during wild-type development. As TGF-α can act both as a membrane-bound and as a diffusible ligand for the EGF-R, *lin-3* may act as the inductive signal in either case. Examination of *lin-3* mutations, *lin-3* expression, and the construction of transgenes with site-directed mutations should confirm whether *lin-3* is the inductive signal, and if so, whether the signal is diffusible.

HYPOTHESIS FOR THE MOLECULAR MECHANISM OF VULVAL INDUCTION

We summarize our current model for the role of *lin-3, let-23*, and *let-60* in the vulval inductive signalling pathway in Figure 9. It is proposed that *lin-3* encodes the inductive signal from the anchor cell and that *let-23* encodes the receptor for the *lin-3* inductive signal. *let-60* would encode a switch protein whose activity state is set by the action of the *let-23* tyrosine kinase. By analogy with mammalian receptor tyrosine kinase-mediated signal transduction pathways (reviewed by Hunter, 1991), *let-23* is expected to control *let-60*

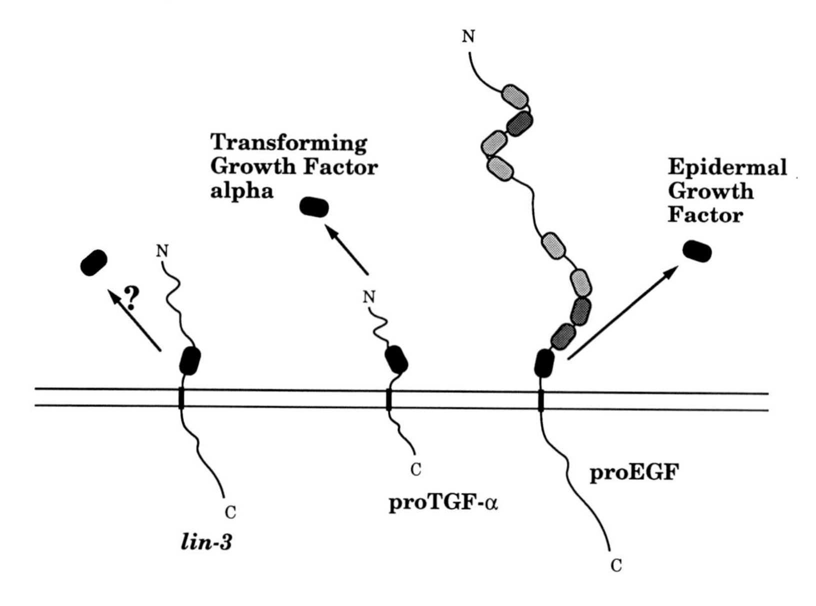

Fig. 8. Schematic of *lin-3* versus other growth factor precursors. TGF-α and EGF are known ligands for the EGF-receptor. Black oval represents the EGF growth factor motif. N=amino-terminus; C=carboxy-terminus.

activity through intermediary proteins, i.e., *let-60* is not necessarily a substrate of *let-23*.

This hypothesis predicts that *lin-3* is expressed in the anchor cell, and that *let-23* and *let-60* are expressed in the VPCs. Experiments to test this hypothesis are in progress. Other genes are likely to act in the vulval induction pathway or as modulators of this pathway. For example the *lin-10* gene is necessary for a full level of vulval induction (Ferguson and Horvitz, 1985; Ferguson et al., 1987; Kim and Horvitz, 1990), although some induction can occur in a putative null mutant of *lin-10* (Sternberg and Horvitz, 1989).

INDUCTION DURING MALE SPICULE DEVELOPMENT

As mentioned above, most mutations in *lin-3, let-23*, and *let-60* are pleiotropic. Loss of function of any of these genes results in lethality during the first larval stage (Herman, 1978; Ferguson and Horvitz, 1985; Clark et al., 1988; Beitel et al., 1990; Han et al., 1990); the cause of lethality is not known. Some mutations in these three genes result in males with crumpled copulatory spicules (Han et al., 1990; Aroian and Sternberg, 1991; R.J.H., unpublished observations). These mutations result in cell lineage defects in the male B cell lineage. Because these genes are involved in inductive signalling during vulval induction, we searched for a source of an inductive signal acting on the B cell progeny during male spicule development.

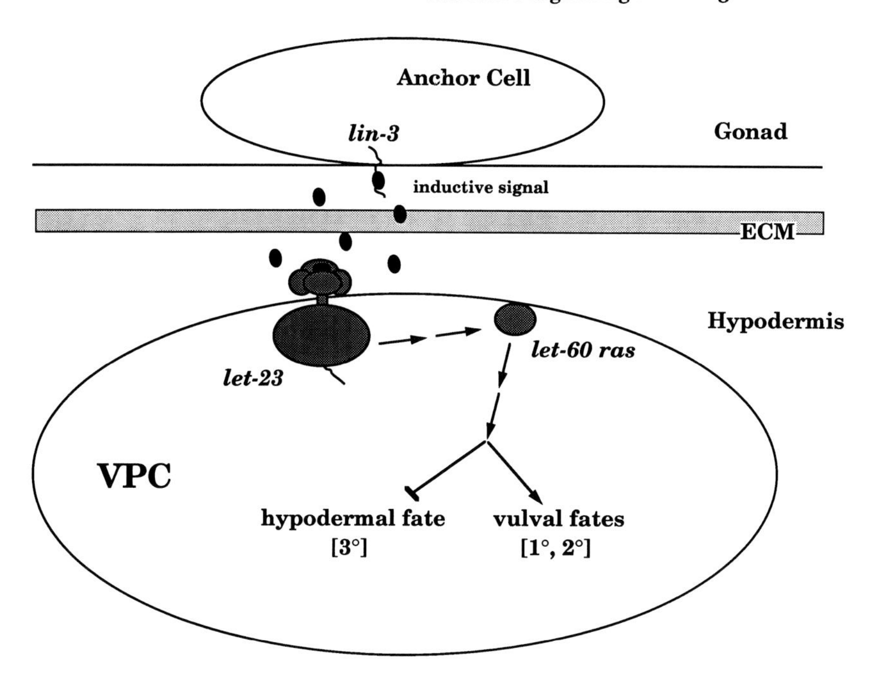

Fig. 9. Hypothesis for the inductive signalling pathway during vulval development. ECM=extracellular matrix.

The cells F and U or their progeny induce some of the B descendants to generate the appropriate set of cells (Fig. 10). By the end of the second larval stage, the male-specific blast cell B had generated 10 descendants (Sulston and Horvitz, 1977). Two of these ten cells, B.alaa and B. araa, are bilaterally symmetric homologs that migrate to the midline and assume unique positions with respect to the anterior-posterior axis of the male larva, one cell being more anterior, its homolog more posterior. The cells B.alaa and B.araa move so that one randomly assumes the more anterior position and thus becomes

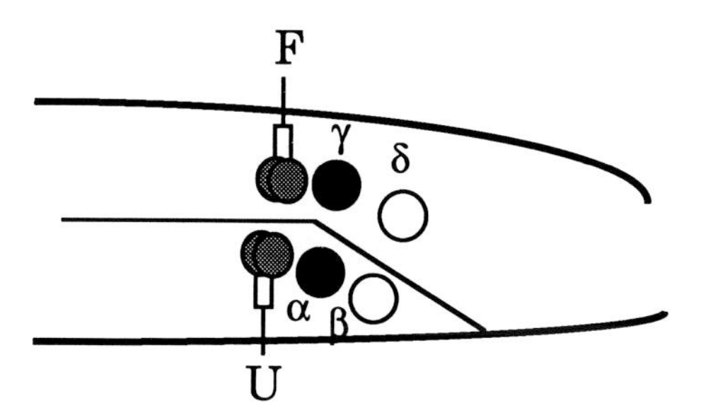

Fig. 10. B, F, U schematic. A schematic lateral view of a young L3 male is shown. Four of the progeny of the B cell are indicated (α, β, γ, δ). At this stage, F and U have divided.

Bα while the other cell assumes the more posterior position and thus becomes Bβ. Bα and Bβ generate distinct types of progeny cells by distinct patterns of cell division. Thus B.alaa and B.araa are equivalent in their developmental potential (i.e., constitute an equivalence group) and their fates depend on positional cues. Similarly, two other homologs, B.alpp and B.arpp, also constitute an equivalence group and migrate (at random) to unique positions (and fates) at the midline. In this group the anterior cell is called Bγ and the posterior cell Bδ, Bγ and Bδ generate distinct types of progeny cells by distinct patterns of cell division.

We have identified some of the positional cues that distinguish Bα from Bβ and Bγ from Bδ. In the absence of both F and U, Bα and Bγ lineages are disrupted. Bα often generates a lineage with characteristics of the Bβ lineage, and Bγ often undergoes fewer rounds of cell division, possibly like the Bδ lineage (H.M.C. and P.W.S., manuscript in preparation). In the absence of U alone, Bα occasionally generates an abnormal lineage. Likewise, in the absence of F alone Bγ occasionally generates an abnormal lineage. Thus, both F and U might produce a signal that can promote the Bα and Bγ fates. F and U are sister cells (Sulston et al., 1983) and both generate very similar sets of progeny cells (Sulston et al., 1980). It is thus not surprising that they might both produce the same signal.

lin-3, let-23, and *let-60* are necessary for Bα and Bγ induction (H.M.C. and P.W.S., manuscript in preparation). In mutant males defective in *lin-3, let-23*, or *let-60*, defects in the B lineage are seen that are similar to the effects of ablation of both F and U. Specifically, Bα often generates a lineage with some characteristics of the Bβ lineage, and Bγ often generates a lineage with some characteristics of the Bδ lineage Thus, *lin-3, let-23*, and *let-60* act during induction of B progeny by F and U. The prediction of this hypothesis is that *lin-3* encodes the F/U inductive signal, and *let-23* and *let-60* act in B.alaa, B.araa, B.alpp, and B.arpp to receive and transduce the inductive signal.

SUMMARY AND PROSPECTS

Our molecular and genetic studies have identified three key genes involved in inductive signalling in two aspects of *C. elegans* development: hermaphrodite vulval development and male spicule development (Fig. 11). These genes encode nematode proteins similar to proto-oncogenes in mammals. The order of action of these genes has been examined during vulval induction using the dominant phenotypes caused by *lin-3* and *let-60* transgenes in tests of epistasis. The *lin-3* gene, which acts first, encodes a growth factor precursor of the EGF family. Whether this precursor is processed to yield a diffusible EGF-like growth factor is under investigation. *lin-3* is thus an excellent candidate for the inductive signal from the anchor cell. The *let-23* gene, which acts after *lin-3* but before *let-60*, encodes a tyrosine kinase of the

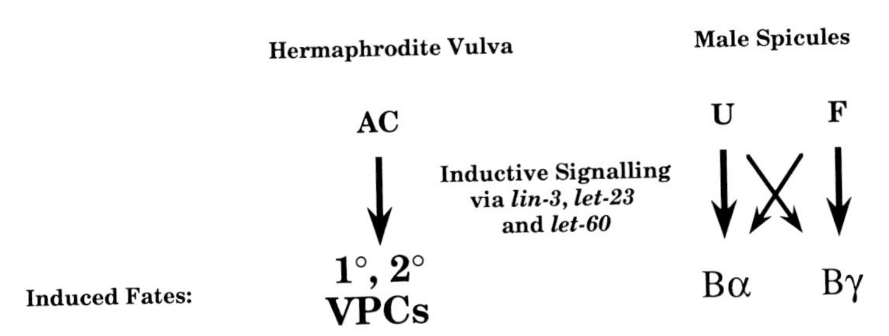

Fig. 11. Parallel between induction of VPCs and B progeny. AC induces 1° and 2° VPCs. F and U or their progeny induce Bα and Bγ.

EGF-receptor family. *let-23* is an excellent candidate for the receptor for the *lin-3* inductive signal. The *let-60* gene acts like a switch whose activity state controls vulval differentiation. *let-60* encodes a *ras* protein, a known "molecular switch." Mutations in the *lin-3, let-23*, and *let-60* genes are pleiotropic, indicating that these genes are involved in multiple aspects of development. By cell ablation experiments, we have found a new example of induction occurring during male spicule development. Like vulval induction, this induction requires the products of *lin-3*, *let-23*, and *let-60*.

Our experiments provide additional examples of intercellular signalling during nematode development. The "piecemeal recruitment" of cells into structures during embryonic development described by Sulston et al. (1983) may be best explained by assuming that cell interactions play a major role in programming the fate of many embryonic cells. We expect that additional cell ablation, genetic, and molecular biological experiments such as those described here will reveal yet more cell interactions in this nematode. For example, if *lin-3* and/or *let-23* are expressed in cells or at times inconsistent with known developmental roles of these genes, this may imply the existence of new intercellular signals.. Also, mosaic analysis of null alleles of either gene might reveal additional roles for these genes that could be masked by the larval lethal phenotype.

Since, as described above, *lin-3*, *let-23*, and *let-60* are involved in two clear cases of inductive signalling, and since tissue-specific mutations of all three genes are known (Herman, 1978; Ferguson and Horvitz, 1985; Clark et al., 1988; Beitel et al., 1990; Han et al., 1990; Aroian and Sternberg, 1991; R.J.H. and P.W.S., unpublished observations), we can address issues of tissue-specific action of intercellular signalling and signal transduction genes. Analysis of the molecular lesions of each tissue-specific mutation might reveal domains or residues of each protein necessary for action in a tissue-specific manner. Genetic analysis, for example, of tissue-specific suppressor mutations, might identify tissue-specific regulators, targets, or cofactors for these genes.

ACKNOWLEDGEMENTS

We especially thank our collaborators in the studies reviewed here: R. Aroian, M. Han, M. Koga, G. Jongeward, J. Mendel, and Y. Ohshima. We thank Andy Golden for insightful editorial suggestions. Many of the strains used in this work were provided by the Caenorhabditis Genetics Center, which is funded by the NIH National Center for Research Resources. Work in our laboratory described here has been supported by the Howard Hughes Medical Institute, the USPHS (HD3690), the March of Dimes Birth Defects Foundation, and the NSF. P.W.S. is an investigator of the Howard Hughes Medical Institute and a Presidential Young Investigator of the National Science Foundation. R. J. H. and H. M. C. have been Predoctoral Fellows of the NSF. H. M. C. is supported by an USPHS training grant and a Helen G. and Arthur McCallum Fellowship.

REFERENCES

Aroian RV, Koga M, Mendel JE, Ohshima Y, Sternberg PW (1990): The *let-23* gene necessary for *Caenorhabditis elegans* vulval induction encodes a tyrosine kinase of the EGF receptor subfamily. Nature 348:693–699.

Aroian RV, Sternberg PW (1991): Multiple functions of *let-23,* a *C. elegans* receptor tyrosine kinase gene required for vulval induction. Genetics 128:251–267.

Barbacid M (1987): *ras* genes. Ann Rev Biochem 56:779–827.

Beitel G, Clark S, Horvitz HR (1990): The *Caenorhabditis elegans ras* gene *let-60* acts as a switch in the pathway of vulval induction. Nature 348:503–509.

Bourne HR, Sanders, DA, McCormick F (1990): The GTPase superfamily: A conserved switch for diverse cell functions. Nature 348:125–132.

Bourne HR, Sanders, DA, McCormick F (1991): The GTPase superfamily: Conserved structure and molecular mechanism. Nature 349:117–127.

Burglin TR, Finney M, Coulson A, Ruvkun G (1989): *Caenorhabditis elegans* has scores of homoeobox-containing genes. Nature 341:239–243.

Carpenter G, Wahl MI (1990): The epidermal growth factor family. In Sporn (ed): "Handbook of Experimental Pharmacology", pp 69–171.

Chalfie M, Horvitz HR, Sulston JE (1981): Mutations that lead to reiterations in the cell lineages of *Caenorhabditis elegans*. Cell 24:59–69.

Clark DV, Rogalski TM, Donati LM, Baillie DL (1988): The *unc-22* (IV) region of *Caenorhabditis elegans*: Genetic analysis of lethal mutations. Genetics 119:345–353.

Coulson A, Waterston R, Kiff J, Sulston J, Kohara Y (1988): Genome linking with yeast artificial chromosomes. Nature 335:184–186.

Coulson AR, Sulston J, Brenner S, Karn J (1986): Toward a physical map of the genome of the nematode *Caenorhabditis elegans*. Proc Natl Acad Sci USA 83:7821–7825.

Ferguson E, Horvitz HR (1985): Identification and characterization of 22 genes that affect the vulval cell lineages of *Caenorhabditis elegans*. Genetics 110:17–72.

Ferguson E, Horvitz HR (1989): The multivulva phenotype of certain *C. elegans* mutants results from defects in two functionally-redundant pathways. Genetics 123:109–121.

Ferguson EL, Sternberg PW, Horvitz HR (1987): A genetic pathway for the specification of the vulval cell lineages of *Caenorhabditis elegans*. Nature 326:259–267.

Finney M, Ruvkun G, Horvitz HR (1988): The *C. elegans* cell lineage and differentiation gene *unc-86* encodes a protein containing a homeo domain and extended sequence similarity to mammalian transcription factors. Cell 55:757–769.

Fire A (1986): Integrative transformation of *Caenorhabditis elegans*. EMBO J 5:2675–2680.
Freyd G, Kim SK, Horvitz HR (1990): Novel cysteine-rich motif and homeodomain in the product of the *Caenorhabditis elegans* cell lineage gene *lin-11*. Nature 344:876–879.
Georgi LL, Albert PS, Riddle DR (1990): *daf-1,* a *C. elegans* gene controlling dauer larva development, encodes a novel receptor protein kinase. Cell 61:635–645.
Greenwald I, Seydoux G (1990): Analysis of gain-of-function mutations of the *lin-12* gene of *Caenorhabditis elegans*. Nature 346:197–199.
Gross RE, Bagchi S, Lu X, Rubin CS (1990): Cloning, characterization, and expression of the gene for the catalytic subunit of cAMP-dependent protein kinase in *Caenorhabditis elegans*. J Biol Chem 265:6896–6907.
Han M, Aroian R, Sternberg PW (1990): The *let-60* locus controls the switch between vulval and non-vulval cell types in *C. elegans*. Genetics 126:899–913.
Han M, Sternberg PW (1990): *let-60,* a gene that specifies cell fates during C. elegans vulval induction, encodes a ras protein. Cell 63:921–931.
Herman RK (1978): Crossover suppressors and balanced recessive lethals in *Caenorhabditis elegans*. Genetics 88:49–65.
Herman RK (1989): Mosaic analysis in the nematode *Caenorhabditis elegans*. J Neurogenetics 5:1–24.
Herman RK, Hedgecock EM (1990): The size of the *C. elegans* vulval primordium is limited by *lin-15* expression in surrounding hypodermis. Nature 348: 169–171.
Hirsh D, Oppenheim D, Klass M (1976): Development of the reproductive system of *Caenorhabditis elegans*. Dev Biol 49:200–219.
Hodgkin J, Horvitz HR, Brenner S (1979): Nondisjunction mutants of the nematode Caenorhabditis elegans. Genetics 91:67–94.
Horvitz HR, Sternberg PW (1991): Multiple intercellular signalling systems control the development of the *C. elegans* vulva. Nature 351:535–541.
Horvitz HR, Sulston JE (1980): Isolation and genetic characterization of cell-lineage mutants of the nematode *Caenorhabditis elegans*. Genetics 96:435–454.
Hu E, Rubin CS (1990): Casein kinase II from *Caenorhabditis elegans*. J Biol Chem 265:5072–5080.
Hunter T (1991): Cooperation between oncogenes. Cell 64:249–270.
Kamb A, Weir M, Rudy B, Varmus H, Kenyon C (1989): Identification of genes from pattern formation, tyrosine kinase, and potassium channel families by DNA amplification. Proc Natl Acad Sci USA 86:4372–4376.
Kim SK, Horvitz HR (1990): The *Caenorhabditis elegans* gene *lin-10* is broadly expressed while required specifically for the determination of vulval cell fates. Genes Dev 4:357–371.
Kimble J (1981): Lineage alterations after ablation of cells in the somatic gonad of *Caenorhabditis elegans*. Dev Biol 87:286–300.
Kimble J, Hirsh D (1979): Post-embryonic cell lineages of the hermaphrodite and male gonads in *Caenorhabditis elegans*. Dev Biol 70:396–417.
Lochrie MA, Mendel JE, Sternberg PW, Simon MI (1991): Homologous and unique G protein alpha subunits in the nematode *Caenorhabditis elegans*. Cell Regul 2:135–154.
Massagué J (1990): Transforming growth factor-α. Biol Chem 265:21393–21396.
Mello CC, Kramer JM, Stinchcomb D, Ambros V (1991): Efficient gene transfer in *C. elegans* after microinjection of DNA into germline cytoplasm: extrachromosomal maintenance and integration of transforming sequences. EMBO J 10:3959–3970.
Priess JR, Thomson JN (1987): Cellular interactions in early *Caenorhabditis elegans* embryos. Cell 48:241–250.
Schnabel R (1991): Cellular interactions involved in the determination of the early *C. elegans* embryo. Mech Dev 34:85–100.
Seydoux G and Greenwald I (1989): Cell autonomy of *lin-12* function in a cell fate decision in *C. elegans*. Cell 57:1237–1245.
Sigurdson DC, Spanier GJ, Herman RK (1984): *Caenorhabditis elegans* deficiency mapping. Genetics 108:331–345.

Sternberg PW (1988): Lateral inhibition during vulval induction in *Caenorhabditis elegans*. Nature 335:551–554.

Sternberg PW, Horvitz HR (1986): Pattern formation during vulval development in *Caenorhabditis elegans*. Cell 44:761–772.

Sternberg PW, Horvitz HR (1989): The combined action of two intercellular signalling pathways specifies three cell fates during vulval induction in *C. elegans*. Cell 58: 679–693.

Stinchcomb DT, Shaw JE, Carr SH, Hirsh D (1985): Extrachromosomal DNA transformation of *Caenorhabditis elegans*. Mol Cell Biol 5:3484–3496.

Sulston JE, Albertson DG, Thomson JN (1980): The *Caenorhabditis elegans* male: Postembryonic development of nongonadal structures. Dev Biol 78:542–576.

Sulston J, Horvitz HR (1977): Postembryonic cell lineages of the nematode *Caenorhabditis elegans*. Dev Biol 56:110–156.

Sulston JE, Horvitz HR (1981): Abnormal cell lineages in mutants of the nematode *Caenorhabditis elegans*. Dev Biol 82:41–55.

Sulston JE, Schierenberg E, White JG, Thomson JN (1983): The embryonic cell lineage of the nematode *Caenorhabditis elegans*. Dev Biol 100:64–119.

Sulston JE, White JG (1980): Regulation and cell autonomy during postembryonic development of *Caenorhabditis elegans*. Dev Biol 78:577–597.

Thomas JH, Stern MJ, Horvitz HR (1990): Cell interactions coordinate the development of the *C. elegans* egg-laying system. Cell 62:1041–1052.

Ullrich A, Schlessinger J (1990): Signal transduction by receptors with tyrosine kinase activity. Cell 61:203–212.

Way J, Chalfie M (1988): *mec-3,* a homeobox-containing gene that specifies differentiation of the touch receptor neurons in *C. elegans*. Cell 54:5–16.

Wood WB (1991): Evidence from reversal of handedness in *C. elegans* for early cell interactions determining cell fates. Nature 349:536–538.

Yochem J, Weston K, Greenwald I (1988): *C. elegans lin-12* encodes a transmembrane protein similar to *Drosophila Notch* and yeast cell cycle gene products. Nature 335:547–550.

Evolutionary Conservation of Developmental Mechanisms, pages 159–184

11. Spatial and Temporal Regulation of *Dictyostelium* Development Through Signal Transduction Pathways

J.A. Powell, G.R. Schniztler, J.H. Hadwiger,
P. Howard, R.K. Esch, A. B. Cubitt, K. Okaichi,
C. Gaskins, S. K. O. Mann, and R. A. Firtel

Department of Biology, Center for Molecular Genetics, University of California,
San Diego, La Jolla, California 92093-0634

INTRODUCTION

The morphogenesis of multicellular organisms is one of the most complex problems in biology, for it requires the understanding of processes at the biochemical, cellular, and multicellular levels. As an organism develops, it must establish polarity and coordinate cell division, cellular adhesion and migration, and cell-type differentiation to ultimately form a properly proportioned, functional organism. Extracellular morphogens (Schaller et al., 1989; Green and Smith, 1990; Summerbell and Maden, 1990) and direct cell–cell communication (Banerjee and Zipursky, 1990; Takeichi, 1990) are often used to establish and refine a pattern of development. These extracellular signals must then be sensed by cells and transduced through a cascade of intracellular biochemical reactions to effect the appropriate changes in cellular movement and differentation. The cellular slime mold *Dictyostelium discoideum* has proven to be a good experimental system for studying many of the common problems of multicellular development. *D. discoideum* cells undergo a well-defined and simple pattern of development, have relatively few potential cell fates, and are haploid and amenable to molecular manipulation.

When nutrients are plentiful, *D. discoideum* cells exist as free-living amoebae. However, when their food source is depleted, the cells initiate a program of multicellular development (Fig. 1). Four to five hours after removal of the food source, a small percentage of the starving cells begin emitting pulses of cAMP. When neighboring cells sense this signal via cAMP receptors expressed on the cell surface (Klein et al., 1988), they move chemotactically towards the source and relay the signal. The molecular basis of chemotaxis is being studied by many labs, and activation of the cAMP receptor has been correlated with both actin nucleation (Hall et al., 1989;

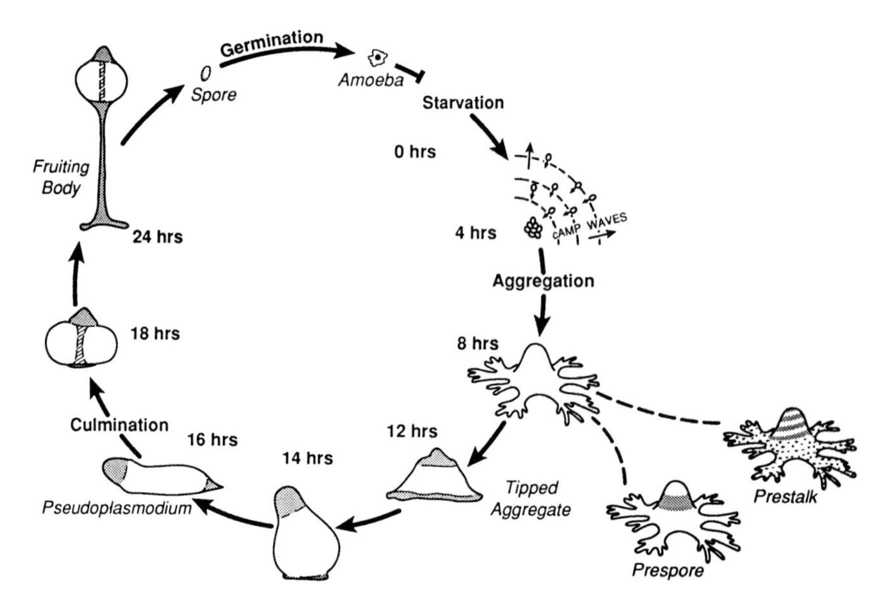

Fig. 1. *Dictyostelium discoideum* asexual life cycle. Starved cells aggregate in response to relayed waves of cAMP, and progress through defined multicellular stages, culminating in a mature fruiting body. Prestalk gene expression (e.g., *Dd-ras*, shown shaded here) is initially localized to a fraction of cells scattered throughout the aggregate. These cells then migrate (a spiral pattern is observed for *Dd-ras*) to form the tip, and part of the base of the mound. Prespore gene expression (e.g., *SP60*, shown shaded here) first appears in cells scattered throughout the mound, but not in the "skirt" cells of the aggregate. As the tip forms, prespore staining cells form a broad "halo" around the tip. In later stages prestalk or stalk regions are shown shaded, while prespore or spore cells are not. For further details, see text.

Dharmawardhane et al., 1989) and changes in myosin phosphorylation and localization (Berlot et al., 1987; Liu and Newell, 1988). Activation of the cAMP receptor also results in activation of adenylate cyclase, which causes the intracellular synthesis and subsequent secretion of cAMP, thus relaying the signal (Janssens and van Haastert, 1987; Firtel et al., 1989). After approximately 1 minute, the system desensitizes, and the cells cease moving and releasing cAMP. This desensitization correlates with the ligand-induced phosphorylation of the cAMP receptors (Vaughan and Deverotes, 1988). Extracellular and cell-surface phosphodiesterases then hydrolyze the cAMP (Podgorski et al., 1988; Franke et al., 1991), the system deadapts, and the cell can respond to the next wave of cAMP. In this way about 10^5 cells are recruited into an aggregation center, and the multicellular organism is formed (see Devreotes, 1989; Firtel et al., 1989, for review).

Pattern formation begins early in development (Figs. 1 and 2) as a tip forms on the mound and then elongates upwards to form a "first finger." The finger

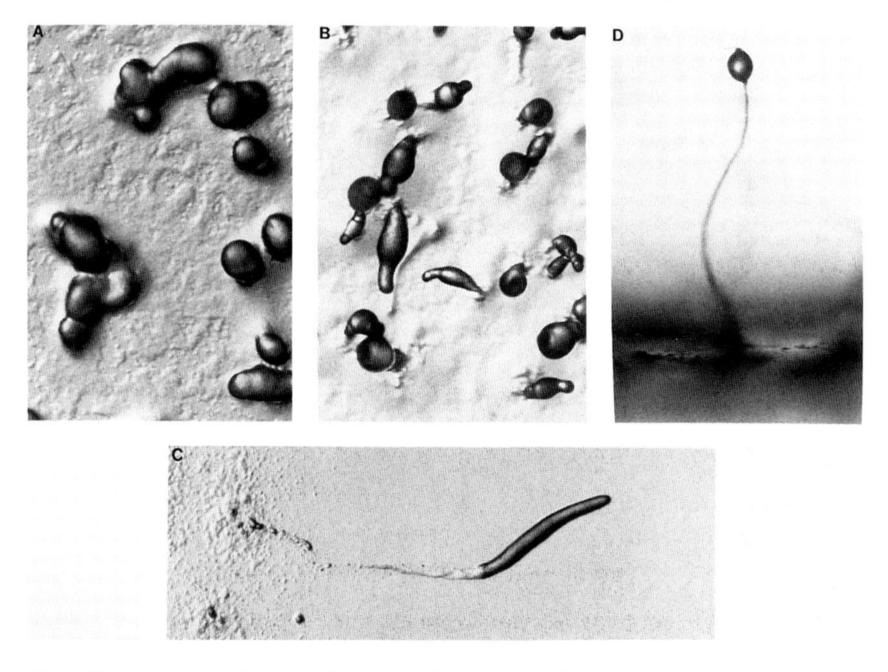

Fig. 2. Photographs of *Dictyostelium discoideum*—multicellular structures. **A**: Mound stage (~9–10 h). **B**: Tipped aggregate (~12h). **C**: Migrating slug or pseudoplasmodium. **D**: Mature fruiting body.

may then fall over to become a phototactic pseudoplasmodium (or slug) which migrates to an area suitable for culmination. The slug forms a "second finger", and as mature stalk cells begin to differentiate, they lift the spore mass about 2 mm above the substratum to form a fruiting body.

D. discoideum differs from many other developmental systems because the multicellular organism is formed by the aggregation of cells. There is no progenitor zygote and, therefore, no maternal influences to establish polarity. Furthermore, the majority of the cells cease division once the aggregate is formed, so cell lineage determinants are not a factor in spatial patterning. *Dictyostelium* development and pattern formation has been studied for decades at the histological and biochemical level (see Loomis, 1982). Recently, molecular techniques have been used to identify some of the cellular components integral to the coordination and execution of morphogenesis. The cloning and identification of these regulatory factors have revealed a surprising degree of evolutionary conservation of developmental strategies. It is now evident that most of the molecular mechanisms used by *D. discoideum* to initiate and guide development are very similar to those found in metazoans. In this chapter, we will review some of the recent progress made in discerning the cellular and molecular mechanisms of *D. discoideum* development and morphogenesis, with emphasis on work accomplished in this lab.

EXTRACELLULAR SIGNALS AND DICTYOSTELIUM MORPHOGENESIS

Cell-type differentiation and pattern formation in *D. discoideum* is thought to be modulated by both cell-autonomous factors and diffusible morphogens within the aggregate. The cell fate of an amoebae can be influenced by its position in the cell cycle when starved (Weijer et al., 1984; Gomer and Firtel, 1987) or by the composition of the food source during vegetative growth (Leach et al., 1973). However, classic slug dissection experiments have shown that cells are not irreversibly committed to their respective fates as they enter the aggregate. In the migrating pseudoplasmodium, about 15% of the anterior and 3% of the most posterior cells express prestalk markers, while the posterior 85%of the organism consists predominantly of prespore cells. Still, the cells are capable of dedifferentiation and redifferentiation, for if the anterior, presumptive prestalk zone of the slug is cut away from the presumptive prespore zone, the two sections will reorganize and eventually form properly proportioned culminants in the absence of cell division (Raper, 1940). Thus, diffusible substances must play a role in the modulation of pattern formation.

When *D. discoideum* cells are starved, they are primed to enter the multicellular phase of their life cycle. However, it is counterproductive for the cells to expend energy in this process if there are too few cells in the immediate vicinity to form a functional multicellular organism. Thus, when starved the amoebae release a density-determining factor (or factors) that must reach a certain extracellular threshold concentration for the cells to initiate the developmental program. One such factor is conditioned medium factor (CMF), a 80-kD secreted glycoprotein (Mehdy and Firtel, 1985; Gomer et al., 1991). Cells starved at very low density do not induce early gene expression and cannot be induced to express late genes in response to high levels of extracellular cAMP (Mann and Firtel, 1989; Gomer et al., 1991). However, adding CMF to the medium bypasses this cell density requirement and allows cAMP-induced gene expression. CMF is thus thought to be a density-dependent extracellular signal which permits the onset of cellular cAMP-relay processes to mediate aggregation.

Other diffusible substances modulate pattern formation once the multicellular organism is formed. At present, the two best characterized potential morphogens are cAMP and differentiation inducing factor (DIF), a chlorinated alkyl phenone (Town et al., 1976; Morris et al., 1987). Adenosine and ammonia have also been identified as diffusible agents that are capable of influencing development (Schaap and Wang, 1986; Feit et al, 1990). Due to the nature of these small molecules, it has been technically difficult to localize them. Therefore, while the effects of these compounds, added exogenously to the organism or to cultured cells, have been studied extensively, their actual roles in development are still subject to much debate (for review see Williams, 1988; Kimmel and Firtel, 1991).

While it is well established that *D. discoideum* amoebae respond to and relay pulsatile cAMP signals during aggregation, the role of cAMP during later development is less well understood. The concentration of extracellular cAMP is thought to reach 1 μM during development (Malkinson and Ashworth, 1972), and most late genes are either induced by high levels of cAMP or require prior exposure to cAMP for their induction (Mehdy et al., 1983; Mehdy and Firtel, 1985; Berks and Kay, 1990). Multiple cAMP receptors (cARs) have been cloned in the laboratories of Kimmel and Deverotes and have been shown to be expressed in developmentally distinct patterns through development (Klein et al., 1988; Saxe, et al., 1991a; see below). Present data indicate that during aggregation, cAMP pulses activate signal transduction pathways through interaction with the cAMP receptor cAR1 and the G protein-containing Gα2 (Kumagai et al., 1989; Sun and Devreotes, 1991; Kumagai et al., 1991).

Differentiation inducing factor (DIF) was first identified as the soluble factor capable of transforming a particularly responsive strain of *D. discoideum* cells to stalk cells in monolayer culture in the presence of cAMP (Town et al., 1976). DIF is first detected during late aggregation and has been shown to induce the prestalk markers *ecm*A and *ecm*B (Williams et al., 1987; Jermyn et al., 1987) and to repress many prespore markers (Early and Williams, 1988; Fosnaugh and Loomis, 1991). A developmentally regulated DIF binding activity has been identified and is being purified in the lab of R. Kay (Insall and Kay, 1990). This activity is present in the cytosol and nucleus of cells and is thus reminiscent of steroid hormone receptors. It is worth noting that while the *ecm*B gene's activation by DIF is inhibited by the presence of cAMP, this activation can only be seen after cells have been exposed to high levels of cAMP in monolayer culture (Berks and Kay, 1988), or after they have been dissociated from late aggregates (Berks and Kay, 1990); these observations indicate that cAMP is still required at an earlier stage to potentiate this gene's expression.

Ammonia is present in relatively high levels in the slug due to the catabolism of proteins. A decrease in ammonia concentration has been shown to induce culmination, and high concentrations of ammonia can increase the expression of prespore genes and promote spore maturation (Schindler and Sussman, 1977; Riley and Barclay, 1990).

Adenosine, which results from the breakdown of cAMP catalyzed by phosphodiesterase and 5′-nucleotidase, may inhibit prespore gene expression by inhibition of cAMP binding to cell surface receptors (Theibert and Deverotes, 1984). Phosphodiesterase and 5′-nucleotidase levels, and thus putative adenosine levels, are highest in the tip (Armant et al., 1980; Brown and Rutherford 1980; Franke et al., 1991), and it has been argued that adenosine may be the tip-released signal that prevents the establishment of other tips and prohibits prespore gene expression in the anterior prestalk zone (Schaap and Wang, 1986).

Direct cell–cell communication may also play a role in morphogenesis, and several developmentally regulated adhesive glycoproteins have been identified (for review see Siu, 1990). Little is known, however, about how different specific contacts may be involved in regulating cell-type differentiation and gene expression, except for the apparent requirement of calcium-dependent contacts for the expression of prespore genes (Mehdy et al., 1983; Mehdy and Firtel, 1985; Fosnaugh and Loomis, 1991).

CELL-TYPE DIFFERENTIATION AND PATTERN FORMATION

Early studies identified two basic cell types in *Dictyostelium* development: prestalk cells and prespore cells. Anterior prestalk cells make up the tip of the migrating slug, and anterior-like cells (ALCs), which have many of the same characteristics as prestalk cells, are scattered throughout the posterior part of the slug. These cell populations eventually become the vacuolated stalk and basal disc cells of the fruiting body and form the upper and lower caps of the sorocarp (Jermyn and Williams, 1991; Esch and Firtel, 1991). The prespore cells differentiate into spores upon culmination (Loomis, 1982; Sternfield and David, 1982). In initial studies, the prespore cells could be readily identified by the presence of prespore vesicles (Takeuchi, 1963), which store spore-coat materials, while anterior prestalk cells and ALCs could be distinguished by staining with neutral red (Bonner et al., 1990). The prestalk cells were also found to be similar to early aggregation cells in their morphology (Schaap, 1983) and their ability to move chemotactically toward a cAMP signal (Mee et al., 1986; Otte et al., 1986). In recent years, we have come to better understand the complexity of *D. discoideum* morphogenesis and pattern formation.

Recent work with antibody and β-galactosidase staining have allowed for the further elaboration of cell types and may also shed some light on how the spatial patterning of the organism is established. These results suggest that the prespore region of the organism is homogeneous since the expression patterns of several genes is virtually identical (Kreft et al., 1984; Gomer et al., 1986b; Williams et al., 1989; Haberstroh and Firtel, 1990; J.A. Powell and R.A. Firtel, unpublished observations), although, as will be discussed later, this homogeneity of expression may be achieved only by the delicately balanced regulation of these genes in response to multiple positional signals. The prespore markers are first expressed during aggregation as a broad halo around the center of the developing aggregate and are not evident in the "skirt" of cells still streaming into the mound. Later, in the migrating slug, uniform staining is seen throughout the posterior 80% of the organism, with a very sharp demarcation between the anterior of the prespore zone and the tip (see Fig. 1 and 3).

The prestalk cell population, in contrast, has been shown to be more complex. This is not unexpected, as the prestalk cells are thought to be responsible for orchestrating many of the more active processes of development. A transplanted tip is capable of creating a second axis and splitting a migrating slug into two organisms, thus acting like an organizing center (Raper, 1940; Rubin and Robertson, 1975). Theories which attempt to explain slug migration argue that prestalk cells are responsible for the energy and direction of motion (Williams et al., 1986). During culmination, prestalk and stalk cells are the driving force for raising the prespore mass off the substrate and forming the fruiting body (see below).

There are at least two populations of prestalk cells as defined by the expression pattern of prestalk-specific genes. The DIF-inducible gene *ecm*A (Jermyn and Williams, 1991) and the cAMP-induced gene *Dd-ras* (Esch and Firtel, 1991) are markers for the prestalk A cells that comprise the majority of the anterior tip (see Fig. 1 and 3). Prestalk B cells, identified by the expression pattern of *ecm*B, stain only a cone of cells in the center of the tip. All of these markers are also expressed in the ALC population that is scattered throughout the rear of the slug. During culmination, ALCs appear to split into three populations: one forming the basal disc upon which the fruiting body rests and two others moving to form the upper and lower cups of cells that surround the spore mass (Jermyn and Williams, 1991; Ceccarelli et al., 1991; Esch and Firtel, 1991).

Analysis of the expression pattern of a *Dd-ras/lacZ* fusion gene early in development provided information about the ontogeny of the prestalk pattern. *Ddras* is expressed in about 10% of growing cells at any one time and the expression diminishes upon the onset of development. In the early aggregate *Ddras/lacZ* expression is detectable in about 10% of the cells, and those cells appear to be randomly distributed. The cells which express *Dd-ras/lacZ* then appear to migrate in a spiral pattern to form the tip, possibly in response to the proposed spiral gradient of cAMP within the aggregate. As development proceeds, staining is evident in anterior prestalk cells, anterior like cells, and stalk cells (see Esch and Firtel, 1991).

To execute this developmental program, *D. discoideum* cells must communicate with each other. As discussed earlier, some of the factors involved in extracellular signalling have been identified, but as yet the mechanisms involved in the intracellular transduction of these signals are not fully understood. To better understand these signal transduction mechanisms, there has been a major effort to identify components of these pathways. Of great interest are the cell surface cAMP receptors. The cAR1 gene encodes one of these receptors and is expressed during the aggregation phase of development (Klein et al., 1988). Since *cAR1* null mutants do not respond to cAMP when starved, this receptor is required for aggregation in early development

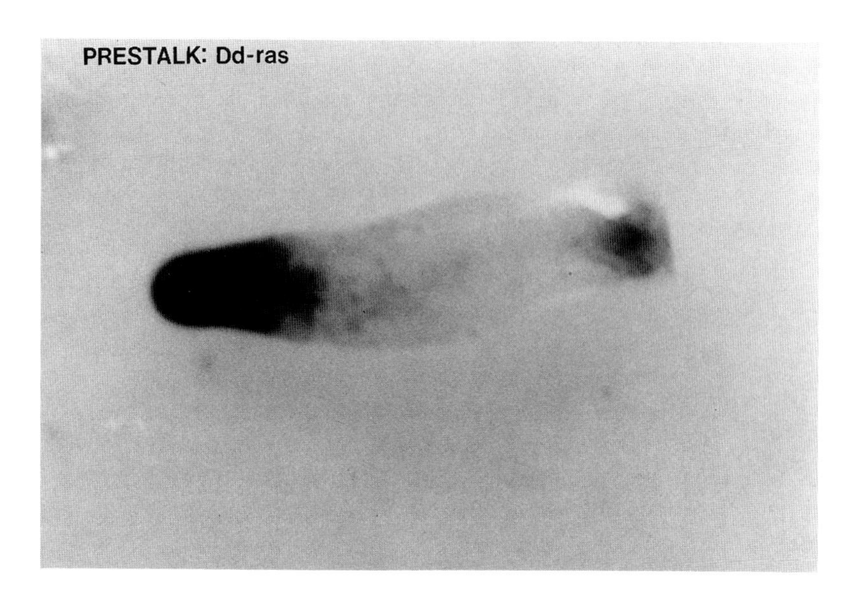

Fig. 3. Cytological staining of *lacZ* transformants at the slug stage of development, with X-gal as the substrate. Dark regions are β-gal positive cells. **Top**: Prestalk-specific staining of a *Ddras/LacZ* transformant slug. **Bottom**: Prespore-specific staining of an SP60/*LacZ* transformant slug.

(Sun and Deverotes, 1991). Genes encoding three homologs of *cAR1 (cAR2, cAR3, cAR4)* have also been cloned. *cAR2* and *cAR4* are expressed during later stages of development, consistent with a role for cAMP signalling in cell differentiation and morphogenesis (Saxe et al., 1991lb; Kimmel and Firtel, 1991). All of the cARs identified thus far have seven transmembrane domains, a characteristic of receptors that are coupled to G proteins.

G PROTEINS AS MEDIATORS OF SECOND MESSENGER PATHWAYS

Heterotrimeric G proteins comprised of α, β, and γ subunits have been found in all eukaryotes examined and are associated with transducing signals from cell surface receptors to intracellular effectors (for review see Simon et al., 1991). Ligand-bound receptors facilitate the exchange of GDP for GTP bound to the α subunit, allowing the α subunit to dissociate from the βγ dimer and to stimulate downstream effectors. Attenuation of this signal results from the α subunit's hydrolysis of GTP and GDP and reassociation with the β and γ subunits. In most eukarotes, many more genes encoding α subunits have been isolated than those for β or γ subunits; this may reflect the total number of genes of each subunit type or, perhaps, the relative ease of isolating and identifying α subunits due to the presence of highly conserved regions thought to be important for guanine nucleotide interactions. Although numerous Gα genes have been identified in many different organisms, very is little is known about the specific function of most of the corresponding Gα subunits. Therefore, classification of G proteins has been primarily based on structure of the Gα subunits. Except for cases of extremely high homology, this classification has had limited application.

There have been at least six distinct Gα genes identified from *Dictyostelium* (Fig. 4), none of which show significant homology with each other or with other known eukaryotic G proteins outside the highly conserved regions (Pupillo et al., 1989; Hadwiger et al., 1991; C. Gaskins and R.A. Firtel, unpublished data; M. Pupillo and P.N. Devreotes, personal communication). Some of these Gα genes, such as Gα1, Gα2, and Gα6, are expressed primarily during vegetative growth and early development, whereas others are expressed primarily during later developmental stages (Gα4 and Gα5). Although the patterns of expression may provide clues to the Gα function, the analysis of null mutants created by gene disruptions offers a genetic approach to understanding Gα function with respect to development processes.

Gα2 expression is induced by cAMP-pulsing during aggregation. Disruption of the Gα2 gene results in cells that are incapable of aggregating and are defective in cAMP receptor-mediated responses, including the activation of adenylate cyclase (AC), guanylate cyclase (GC), phospholipase C (PLC), developmental gene expression, and chemotaxis to cAMP (Kumagai et al.,

DEVELOPMENTAL REGULATION OF
Gα SUBUNITS AND cAMP RECEPTORS

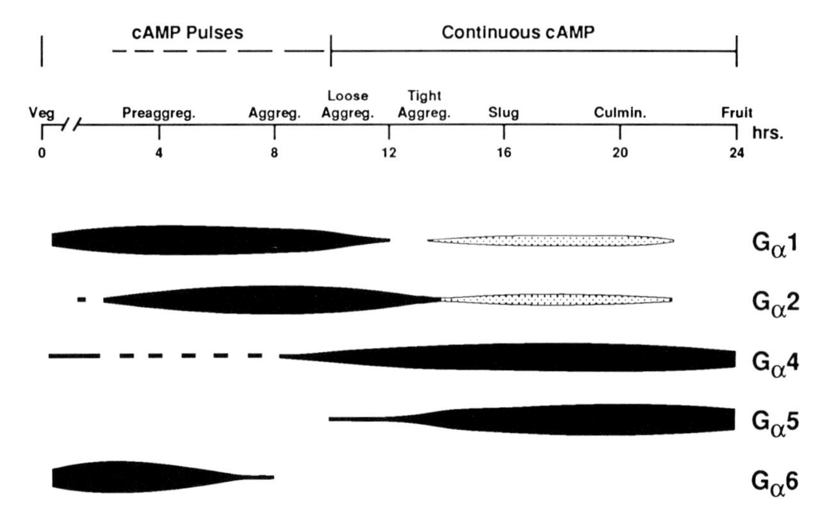

Fig. 4. The expression pattern of Gα genes during development. Presented are the temporal expression patterns (determined by Northern analysis) of the Gα genes described in this chapter. The thickness of the bars reflects the relative levels of expression. Also indicated are the morphological stages and the extracellular cAMP patterns during the *D. discoideum* life cycle. Cyclic AMP pulses initiate a few hours after the onset of starvation, and as the cells form a multicellular organism, extracellular cAMP levels increase and are more continuous.

1991; Okaichi et al., submitted for publication). These phenotypes are similar to those of *Frigid*A mutants (Coukell et al., 1983, Kesbeke et al., 1988) that have been shown to contain mutations in the Gα2 locus (Kumagai et al., 1989). In addition, membranes from gα2 null cells show no high-affinity binding sites for extracellular cAMP, suggesting that the Gα2-containing G protein is directly coupled to cAR1, the cAMP cell surface receptor which is preferentially expressed during aggregation (Klein et al., 1988). Biochemical analysis of gα2 null mutants and strains expressing mutant Gα2 proteins suggests that Gα2 directly couples to PLC and possibly to GC; however, Gα2 does not appear to be the Gα subunit that directly activates AC. While gα2 null mutants do not show cAMP-activation of AC in vivo, AC can be activated in vivo by GTPγS (Kesbeke et al., 1988; Snaar-Jagalska and van Haastert, 1988). A current model of Gα2 function is illustrated in Figure 5. In addition to the requirement of Gα2 for aggregation, there may be a functional role for Gα2 later in development, during culmination, as suggested by prestalk specific expression of the Gα2 gene during this period (F. Carrel and R.A. Firtel, unpublished observations).

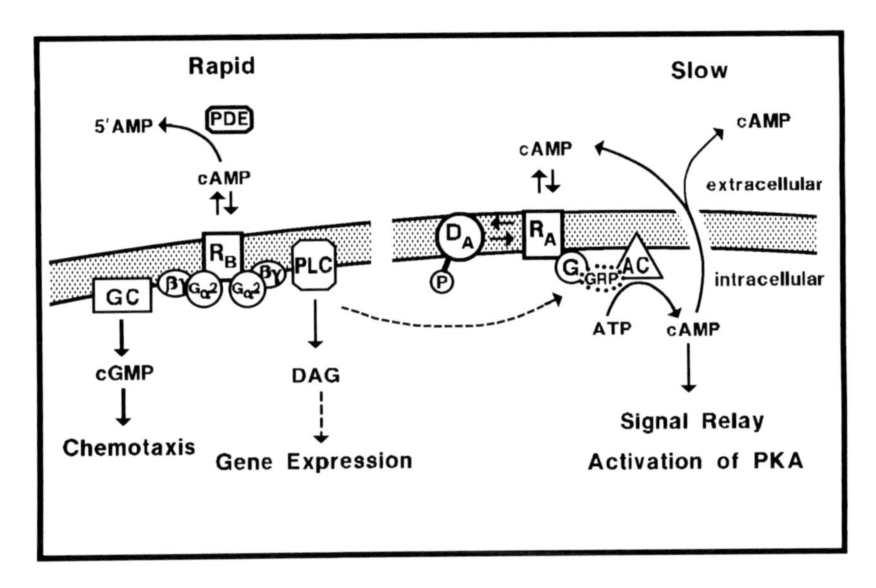

Fig. 5. Model for cAMP-receptor-mediated signal transduction pathways active during early development. The responses to extracellular cAMP are divided into a slow pathway on the right that involves the activation of adenylate cyclase (AC), and the rapid response pathways that include the activation of guanylate cyclase (GC). Activation of adenylate cyclase is mediated by the receptor and a G protein and requires the soluble factor GRP, which is deficient in the *synag* 7 mutants (Theibert and Devreotes, 1986). Cyclic AMP is secreted and/or released from the cells into the extracellular medium where it can activate unoccupied receptors on the same cell or activate receptors on adjacent cells, thus relaying the signal during aggregation. The receptor is shown to be phosphorylated in response to ligand binding, and this modification is thought to result in adaptation of the receptor. Activation of cAMP-dependent protein kinase (PKA) is shown as an intracellular function of cAMP, for PKA is thought to modulate developmental responses and gene expression. The rapid response pathways induce chemotaxis and regulate gene expression. Cyclic AMP binds to receptors that are coupled to a heterotrimeric G protein that includes the subunit Gα2. PLC and GC are shown as possible direct effectors of Gα2. Chemotaxis is thought to involve the activation of guanylate cyclase and the production of cGMP. Extracellular cAMP-phosphodiesterases metabolize the cAMP, thus allowing the receptors to de-adapt.

To further analyze Gα2 function, missense mutations were made in Gα2 and expressed in gα2 null cells and in wild-type cells to examine both recessive and dominant phenotypes. Two of these mutations, a G(40)−>V at the second G in the GAGES box and a Q(208)−>L in the GGQRS domain, were expected to be "dominant activating" mutations, since they are believed to reduce the inherent GTPase activity in Gα subunits and thus increase the fraction of the active, GTP-bound Gα subunit in the absence of receptor activation (Gupta et al., 1991). Unexpectedly, expression of the Q(208)−>L or G(40)−>V construct in gα2 null cells does not restore cAMP-mediated activation of AC, GC, PLC, or pulse-induced gene expression. Overexpression of either construct in wild-type cells results in an unexpected dominant negative phenotype (Okaichi et al., 1992). The cells aggregate poorly, and there is an almost complete inhibition of receptor-mediated responses.

The only other Gα gene expressed early in development that has been analyzed is the Gα1 gene. Since deletion of this gene does not result in any detectable phenotype with respect to vegetative growth or development, Gα1 function is probably not essential for these processes, or this function maybe encoded by multiple genes (Kumagai et al., 1991). Overexpression of the Gα1 gene results in multinucleated cells during vegetative growth (Kumagai et al., 1989). However, this phenotype may or may not have a direct relationship to Gα1 function.

ROLE OF G PROTEINS IN LATER DEVELOPMENT

As mentioned earlier, the fact there are at least two G proteins, Gα4 and Gα5, that are expressed during these later stages of development implies that G protein-mediated signal transduction pathways may function in cell differentiation and morphogenesis (Hadwiger et al., 1991) The Gα4 gene is primarily expressed after the aggregate has formed, and this expression appears to be most prominent in a small subset of cells distributed over the entire multicellular organism with a pattern similar to that of anterior-like cells. The Gα4 subunit contains some unusual sequence divergences in regions highly conserved in most eukaryotic Gα subunits. These divergences, such as the substitution of a proline for an alanine residue in the highly conserved GAGES box, are located within regions that are thought to interact with guanine nucleotide, suggesting that Gα4 might represent a novel class of Gα subunits. The GPGES box of Gα4 has also been found in a human Gα subunit, Gα16 (expressed in hematopoietic cells), indicating that this sequence divergence is not restricted to *Dictyostelium* (Amatruda et al., 1991). However, the homology between the Gα4 and Gα16 subunits is limited to the highly conserved regions that are important for guanine nucleotide interactions. Therefore, these subunits may function in completely different types of signal transduction pathways.

Gene disruption of the Gα4 gene results in cells that show reduced levels of developmental gene expression, loss of spore production, and abnormal developmental morphology (Hadwiger and Firtel, 1992). These cells aggregate normally but, at the first-finger stage, continue to extend apically to produce long protrusions; at this point development arrests. This phenotype can be complemented by expressing low copy numbers of a wild-type Gα4 vector construct. However, overexpression of the Gα4 gene (Gα4^{HC}) also results in cells that undergo an aberrant morphogenesis distinct from Gα4 null cells and are also deficient in spore production. These cells show delayed formation of the mound and produce a terminal phenotype that contains a "club-like" structure on top of a thick stalk. The phenotypes of gα4 null and Gα4^{HC} cells can be partially rescued in chimeras by the presence of wild-type cells, suggesting that the mutants are deficient in the intercellular signalling required for proper multicellular development. The Gα4 subunit also appears to function in a cell-autonomous manner with regard to the expression of some genes (i.e., *Ddras*) and spore production. Although the Gα4 gene is not expressed prominently in prespore cells, it is possible that a very low level of Gα4 expression in these cells might be necessary for the cell-autonomous function. Alternatively, during vegetative growth or early development, Gα4 function might affect spore production by regulating parameters such as the ratio of cells predisposed to form prespore or prestalk cells.

At present less is known about the Gα5 gene. This gene is primarily expressed during and after the tipped-mound stage of development (Hadwiger et al., 1991). This temporal expression is similar to that of some prespore genes. The receptors or downstream effectors that are coupled to the Gα4 or Gα5 subunits are not known. However, potential candidates include the homologs to the cARI receptor which are expressed, like Gα4 and Gα5, during the multicellular stages of development.

DICTYOSTELIUM RAS

Two very similar *ras* genes with distinct patterns of expression have been identified in *D. discoideum. DdrasG* is expressed primarily during vegetative growth (Robbins et al., 1986). *Ddras* is expressed at low levels in growing cells, is not expressed between the onset of development and aggregation, and is then reexpressed during the multicellular stages (Reymond et al., 1984). The *Dictyostelium ras* proteins show 82% amino acid identity relative to one another, and each is between 60% and 69% conserved relative to the human, *Drosophila*, and yeast *ras* proteins (Robbins et al., 1989). Expression of an activated *Ddras* G(12)$->$T, produces an abnormal developmental phenotype characterized by the formation of multiply tipped aggregates and inhibition of further development (Reymond et al., 1986). Consistent with *Ddras* expression being induced in the early multicellular structure, a great increase in *Ddras* transcripts is observed when cells are treated with constant high levels of cAMP (Reymond et al., 1984). In situ β-galactosidase staining of transformants expressing *lacZ* driven from the *Ddras* promoter (*Ddras/lacZ*) has been used to monitor *Ddras* expression through development, and we have shown *Ddras* expression to be prestalk-specific (Esch and Firtel, 1991). The presence of *Ddras* specifically in the prestalk, tip-forming cells is consistent with the abnormal developmental phenotype resulting from the expression of the activated *Ddras* Thr12 and suggests an essential role for *Ddras* in governing spatial differentiation.

KINASES AND PHOSPHATASES

In other systems many, and perhaps most, aspects of growth regulation are modulated by a complex interplay of protein kinases and phosphatases. This unusually large and rapidly expanding protein family now contains over 100 members (Hunter, 1987). When phylogenetic trees are drawn using the catalytic domain, subfamilies emerge that reveal a clustering of kinases with similar properties, e.g. the protein-tyrosine kinase subfamily (PTK) and the cyclic nucleotide-dependent (PKA) and protein kinase C (PKC) subfamilies (Hanks and Quinn, 1991). This clustering allows both placement of novel kinases within the family, using sequence alignment scoring, and the opportunity to design probes for isolating particular subfamily kinases.

Dictyostelium researchers have begun to address the role these kinases might play in directing the development of *Dictyostelium*. Given the prominent roles that cAMP plays in this organism both as a chemoattractant and as an inducer of cellular differentiation, a kinase targeted for study has been the cAMP-dependent protein kinase. The holoenzyme in *Dictyostelium* is composed of single regulatory (R) and catalytic (C) subunits as compared with the typical R_2C_2 structure found in other organisms (Majerfeld et al., 1984; Mutzel et al., 1987). Researchers from the labs of M. Veron and J. Williams showed that overexpression of the regulatory subunit results in a failure of cells to aggregate (Simon et al., 1989). Presumably, overproduction of the R subunit results in repression of the kinase activity of the catalytic subunit, indicating a role for cAMP-dependent protein kinase signalling in development. Interestingly, when wild-type or mutant forms (that lack both sites for cAMP binding) of the mouse type I R subunit are overexpressed in *D. discoideum*, a similar phenotype is observed. In these cells, activation of guanylate and adenylate cyclases is normal, as is cAMP-pulse-induced gene expression (Firtel and Chapman, 1990). Elaboration of these results awaits the isolation of the gene encoding the catalytic subunit.

Exploiting the homology of the catalytic domain, several labs have begun to isolate *Dictyostelium* members of the kinase superfamily using PCR technology. Over 14 kinases have thus far been identified, some demonstrating developmental regulation (Haribabu and Dottin, 1991; Mann and Firtel, 1991). One of these developmentally regulated kinases, a member of the protein serine/threonine subfamily (by homology), is required for development (Mann and Firtel, 1991). When this gene, *DdPK3*, is disrupted by homologous recombination, cells fail to aggregate. If *Ddpk3* null cells are mixed with wild-type cells, they coaggregate but are left as a discrete mass as the wild-type cells continue through development. Activation of adenylate cyclase is normal in these cells, but prespore genes and late *cAR1* transcripts are not induced.

Developmentally regulated changes in tyrosine phosphorylation have been observed in *Dictyostelium* (Schweiger et al., 1990; P. Howard, B.M. Sefton, and R.A. Firtel, unpublished observations), prompting investigators to isolate protein tyrosine kinases (PTKs). Two developmentally induced PTKs have been isolated by screening expression libraries with antibodies against phosphotyrosine (Tan and Spudich, 1990).

With striking examples of protein tyrosine phosphatases (PTPs) playing primary roles in signal transduction (for review see Fischer et al., 1991), we were encouraged to investigate members of this newly emerging and rapidly expanding protein family. Serendipitously, a gene fragment already cloned in a screen for developmentally regulated genes was identified as a PTP (*PTPD1*). Two more were identified in a PCR screen (*PTPD2, PTPD3*) (P. Howard, C. Allen, and R.A. Firtel, unpublished observations). All three genes possess low levels of vegetative expression and are induced with the onset of development. PTPD1 has been the most intensively studied to date. It contains a domain having about 40% amino acid identity with the conserved catalytic domain I of mammalian and *Drosophila* receptor-like PTPases, and encodes a protein of 68 kD, as expected. The PTPase domain is found at the carboxy-terminus, and there is a consensus site for myristolation at the amino-terminus. If this myristolation site is employed to anchor the protein at the plasma membrane, this would represent a new structural subclass of PTPases. Messages of 1.8 kb and 2.4 kb induced with development are expressed at low levels throughout development. Additional messages of 2.0 kb and 2.2 kb appear at approximately 10 h of development. To examine cell-type expression, a construct consisting of about 1 kb of upstream *PTPD1* DNA and 11 codons of *PTPD1* coding sequence fused in frame to *lacZ* was stably introduced into *Dictyostelium*. Detectable staining first occurs after 9 h of development, with the staining cells localized at the base of the aggregate. Staining is later evident in the tip of the aggregate and subsequently in the slug tip. Remarkably, as the organism approaches culmination, the staining pattern moves to the posterior of the slug and is finally associated with the upper and lower caps of the spore mass. To study PTPD1 function, a loss of function mutant was generated by targeted gene disruption, using the auxotrophic selection *Thy1*. The resulting *ptpD1* null mutants develop normally but are accelerated by 2–3 hrs. When the gene is overexpressed behind its own promoter, the cells exhibit slow growth and unusual developmental morphology, suggesting an important role for protein tyrosine phosphorylation in regulating *Dictyostelium* development.

REGULATION OF DIFFERENTIAL GENE EXPRESSION VIA SIGNAL TRANSDUCTION PATHWAYS

Although we have identified many of the intracellular signalling components involved in executing the developmental program and now understand some of their interactions, we have not yet delineated the complete pathways that eventually lead to changes in gene expression. Towards that end, we are also studying promoters from several classes of cAMP-induced genes to identify important *cis*-acting elements and the corresponding *trans*-acting factors.

The cAMP-pulse-repressed genes are induced upon the onset of development. As the cells respond to the waves of cAMP emanating from the aggregation center, the "pulse-repressed" genes cease to be transcribed (Mann et al., 1988), and a class of "pulse-induced" genes is activated (Mann and Firtel, 1987). Some of the genes induced by the nanomolar pulses of cAMP, such as those encoding cAR1, Gα2, the serine esterase D2, and the cell adhesion molecule, contact sites A, are important for aggregation and early development (Gerisch et al., 1975; Mann and Firtel, 1987; Klein et al., 1988; Kumagai et al., 1989; Rubino et al., 1989; Kumagai et al., 1991).

*Frigid*A mutants, which carry *Gα2* mutations, and *gα2* null strains, do not activate pulse-induced genes in response to cAMP and overexpress pulse-repressed genes. Conditions that result in the continued adaptation and down regulation of cAR1 receptors such as continuous, high levels of cAMP give the same effects (Mann et al., 1988). Thus, this early transduction cascade requires cAR1/Gα2-mediated, signalling processes. Mutant studies have indicated that neither adenylate cyclase (Mann et al., 1988) nor protein kinase A (Firtel and Chapman, 1990) activation appear to be necessary for pulse-induced gene activation. Moreover, mutants that affect the level of cGMP production do not alter the expression of these genes (Mann et al., 1988). It has therefore been postulated that phospholipase C activation is a potential regulator of pulse-induced gene expression. Other classes of early genes are probably induced by different regulatory pathways. Kimmel and coworkers have shown that the pulse-repressed gene *M4-1*, which is maximally expressed during vegetative growth, requires cAMP receptor activation of AC for the repression of gene expression, suggesting that PKA may play a role in this pathway (Kimmel and Carlislile, 1986). Moreover, the aggregation-stage phosphodiesterase promoter and the gene encoding α-fucosidase are induced by both pulses and continuous levels of cAMP (May et al., 1991; Franke et al., 1991).

Later in development, a new complement of cAMP receptors and Gα proteins are expressed (Saxe et al., 1991b; Hadwiger et al., 1991). The levels of extracellular cAMP are much higher and more constant than during aggregation (Mee et al., 1988; Otte et al., 1988). cAMP-inducible late genes are activated by high, continuous levels of cAMP, and regulation of late genes is expected to be much more complex and, in many ways, qualitatively different from that of early genes. Late-gene activation by cAMP has been extensively studied at the level of the activated promoters themselves to find the *cis*-sequences which control this induction, and the proteins that bind them.

The *SP60* prespore promoter is perhaps the best understood late-gene promoter with regard to its cAMP regulation. Analysis of this promoter resulted in the identification of three CA-rich elements (CAEs) that are necessary for cAMP and developmental activation of *SP60* expression. These CAEs are highly homologous to each other and to elements found in other prespore gene promoters (Fosnaugh and Loomis, 1991; J.A. Powell and R.A. Firtel, unpublished observations). Deletion of any one of these elements results in a 10- to 50-fold decrease in expression levels, while a 5′ deletion of all three elements results in no detectable expression. Experiments using *SP60/lacZ* fusion constructs showed that the full promoter directed uniform expression within and restricted to, the prespore zone. However, if the most distal (5′-most) CAE is deleted or mutated, an anterior-to-posterior gradient of β-gal staining is seen within the prespore zone of the slug. In contrast, mutation of the most proximal CAE results in a reverse gradient of expression in transformants (posterior-to-anterior). Mutation of the middle CAE, however, does not affect the spatial distribution of the fusion protein (see Fig. 6). These results are suggestive of previously unsuspected gradients of morphogens or other factors within in the prespore zone that somehow interact with the CA elements in the *SP60* promoter (Haberstroh and Firtel, 1990; Haberstroh et al., 1991).

In gel shifts, two nuclear factors have been shown to bind the CAEs specifically, but with different relative affinities. A developmentally regulated, cAMP-induced binding activity (designated "B") binds most strongly to the two distal (5′-most) CAEs, while a constitutively expressed activity

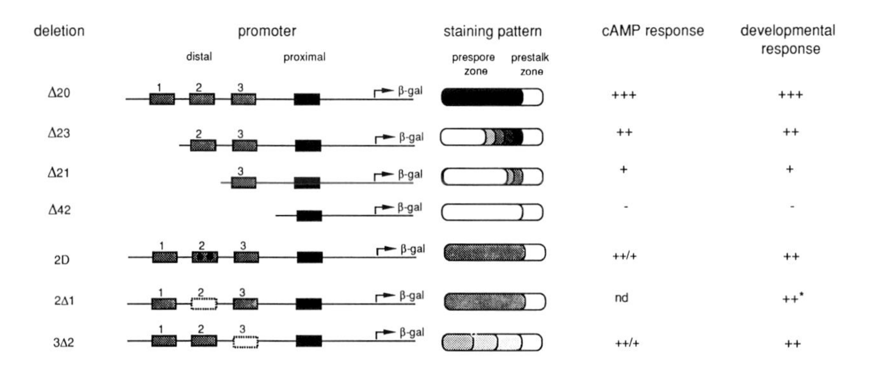

Fig. 6. *SP60* promoter mutations and resultant expression patterns. Shown are schematic representations of mutations made in the *SP60* promoter and the staining patterns seen when these promoters are used to drive expression of a *Sp60/lacZ* fusion gene. The relative expression from mutant *SP60*/luciferase vectors during development or in response to cAMP in shaking culture is also shown. Construct Δ20 contains the full-length promoter, and Δ23, Δ21, and Δ42 are 5 ′ deletion mutants. Constructs 2D, 2Δ1, and 3Δ2 contain site-directed mutations of the promoter. This data has been described in detail previously (Haberstroh and Firtel, 1990; Haberstroh et al., 1991).

(designated "A") (perhaps composed of many distinct complexes) binds most strongly to the proximal CAE (Fig. 7; Haberstroh et al., 1991). The in vitro binding data taken together with the promoter analysis are suggestive of models that include either opposing gradients of the two activators or overlapping gradients of a "B" activator and a repressor (Fig. 8). How such gradients are established is unknown, although we would expect that cAMP has a major role in creating them.

Preliminary work in this lab has shown that the two CAEs in the *14-E6* prespore gene are also critical for developmental induction, and, in vitro, both are capable of binding specific activities similar to "B" and "A" above (J.A. Powell and R.A. Firtel, unpublished observations). Thus, cAMP up-regulation via the binding of factors to CAEs may be a common theme for the tightly coordinately regulated prespore genes.

What of the other late genes which are cAMP activated? Largely heterogeneous in their timing and location of expression, the only property that these genes all share is a late activation by high levels of cAMP. The first, and most extensively studied of these promoters is that of the prestalk-enriched gene *pst-cath/CP2*. Early work identified a G-box sequence which was required for the cAMP induction of the gene (Pears and Williams, 1987; Datta and Firtel, 1987). Over the years many, somewhat similar G-boxes from a variety of other late-gene promoters have been shown to substitute for the *pst-cath/CP2* G-box in vivo, whereas specific mutations in the *pst-cath/CP2* box could not (Pears and Williams, 1988; Hjorth et al., 1990). The developmentally regulated G-box binding factor (GBF) was shown to bind to the *pst-cath/CP2* G-box specifically, as well as to G-boxes from other genes and its ability to bind in vitro correlated with the ability of these sequences to restore activity in vivo (Hjorth et al., 1989; Hjorth et al., 1990). Binding to the *pst-cath/CP2* G-box is cooperative, requiring two DNA half-sites, and recent work indicates that GBF binds either as a homodimer, or a heterodimer of very similarly sized proteins (G.R. Schnitzler and R.A. Firtel, unpublished observations).

While apparently dissimilar, the *pst-cath/CP2* G-box and the *SP60* CAEs both compete for GBF binding. Thousandfold purified GBF fractions bind to the *pst-cath/CP2* G-box as well as to CAEs from both *SP60* and *14-E6*, and the CAE "B" activity and GBF have very similar competition profiles (although certain solution binding requirements differ). Furthermore, the *SP60* CAEs have been recently shown to functionally replace the *pst-cath/CP2* G-box in vivo (C. Gaskins and R.A. Firtel, unpublished observations). These results show that GBF and the "B" activity may either share a component or be identical. They also indicate that CAEs and G-boxes may be cell-type nonspecific late-gene enhancers, and that other promoter sequences are left to determine cell type. In this model, GBF is a *trans*-acting factor whose expression and activity is directly mediated via cAMP-activated signalling pathways. However, the localization of GBF in the organism is unknown, and, in gel-shifts, there are many other specific activities that

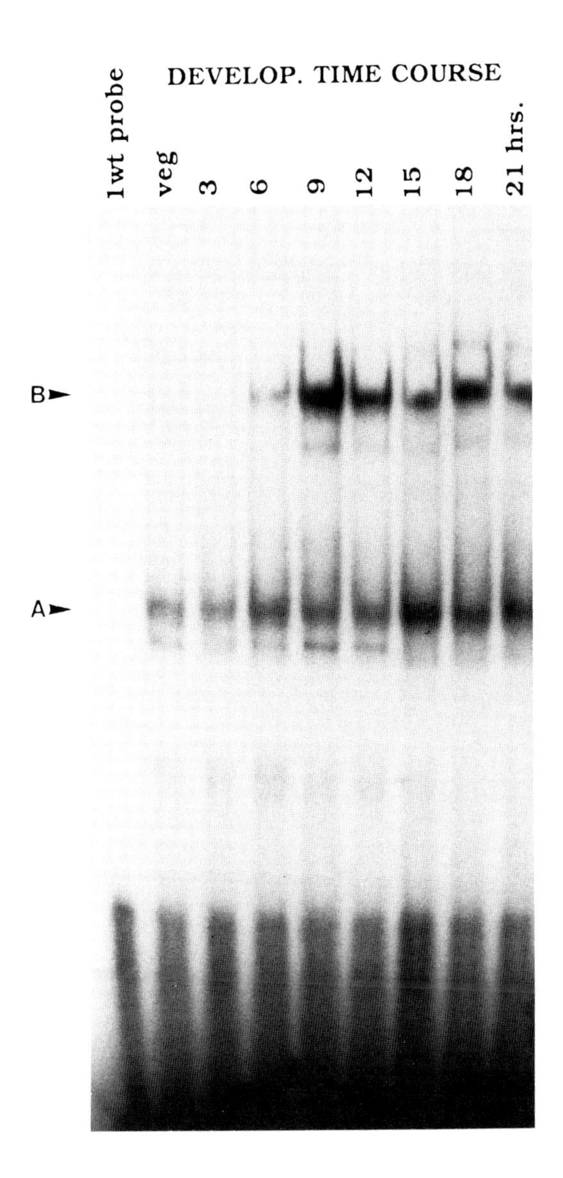

Fig. 7. Developmental time course of CAE1-binding proteins. Nuclear extracts were made at varying times of development (vegetative through 21 hours), were incubated with a CAE1-containing oligonucleotide in the presence of nonspecific competitor, and were electrophoresed in a mobility shift assay. (For details see Haberstroh et al., 1991.) The slower mobility complex (labelled "B") is formed only in extracts from cells in multicellular development, while the faster mobility complex (labelled "A") is formed in all the extracts.

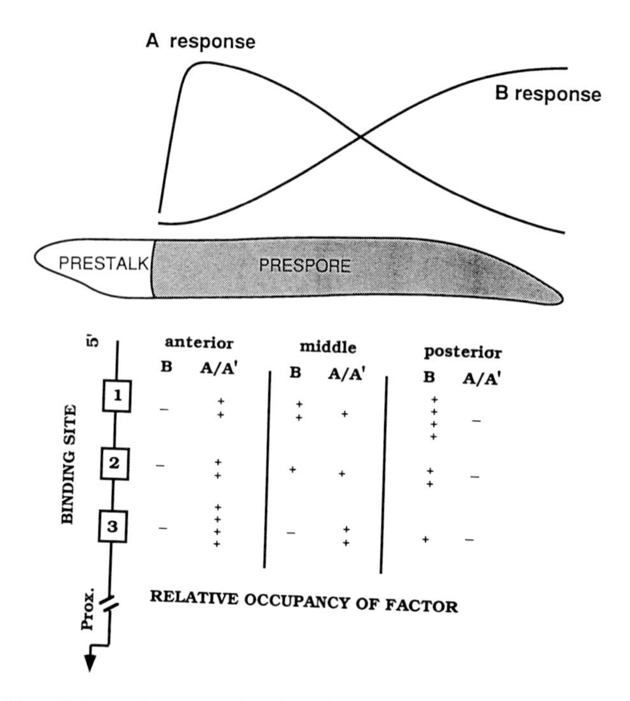

Fig. 8. Model for *SP60* spatial expression: B and A/A′ are hypothesized to be transcriptional activators of equal potency. *SP60* CAE3 has the highest relative affinity for A/A′ and the lowest for B, while CAE1 has the highest affinity for B and moderate affinity for A/A′. CAE2 has a moderate relative affinity for both factors. Activation of the promoter is dependent upon the overall occupancy of these sites by A/A′ and B ("++++" = 100% occupied, "−" = 0%). In the full promoter overlapping gradients of A/A′ and B ensure the same level of activation throughout the slug. The model can explain the observed β-gal staining patterns. In an *SP60* promoter lacking CAE1, most of the activation in the posterior has been lost without affecting activation in the anterior prespore zone, as a result the gradient of expression is strongest in the anterior. The reverse effect is seen when CAE3 is deleted, while deletion of CAE2 results in an even, although lowered, level of activation. The hypothesized gradients may be due to transcription factor transcription, translation, and/or modification by a wide variety of possible extracellular signals.

interact only with G-boxes or only with CAEs, leaving open the possibility that the slightly different complexes of tissue-specific factors that assemble on these sequences may lead to differential expression. In support of this, it has recently been shown that one of the *SP60* CAEs has the ability to confer a prespore expression pattern to the *pst-cath/CP2* G-box-deleted promoter (C. Gaskins and R.A. Firtel, unpublished observations). Work is in progress to clone GBF and the "B" activity, and to further characterize "A," and other G-box-specific activities, and these results should give us further insight into the in vivo function of these proteins.

CONCLUSIONS

Although we have only just begun to explore the detailed molecular interactions involved during signal transduction in *Dictyostelium discoideum*, it seems clear that many aspects of the basic signal transduction mechanisms in *Dictyostelium* are similar to those in a wide variety of other organisms. However, as our detailed understanding of the signal transduction pathways increases, it is becoming apparent that these processes have evolved in *Dictyostelium* in order to suit the particular constraints imposed by its lifestyle and developmental program. Further molecular genetic analysis of the components of these pathways, combined with analysis of the specific mechanisms for controlling the spatial localization of gene expression, should allow us to better understand multicellular development in this relatively simple eukaryotic system.

ACKNOWLEDGEMENTS

Preparation of this manuscript and the work presented from this laboratory was supported by USPHS grants to R. A. F.

REFERENCES

Amatruda TT III, Steele DA, Slepak, VZ, Simon MI (1991): G α 16, a G protein α subunit specifically expressed in hematopoietic cells. Proc Natl Acad Sci USA 88:5587–91.

Armant DC, Stetler DA, Rutherford CL (1980): Cell surface localization of 5′ AMP nucleotidase in prestalk cells of *Dictyostelium discoideum* J Cell Sci 45:119–129.

Banerjee U, Zipursky SL (1990): The role of cell-cell interaction in the development of the *Drosophila* visual system. Neuron 4:177–187.

Berks M, Kay RR (1988): Cyclic AMP is an inhibitor of stalk cell differentiation in *Dictyostelium discoideum*. Dev Biol 126:108–114.

Berks M, Kay RR (1990): Combinatorial control of cell differentiation by cAMP and DIF-1 during development of *Dictyostelium discoideum*. Development 110:977–984.

Berlot CH, Deverotes PN, Spudich JA (1987): Chemoattractant-elicited increases in *Dictyostelium* myosin phosphorylation are due to changes in myosin localization and increases in kinase activity. J Biol Chem 262:3918–3926

Bonner JT, Feit IN, Salassie AK, Suthers HB (1990): Timing of the formation of the prestalk and prespore zones in *Dictyostelium discoideum*. Dev Genet 11:439–441.

Brown SS, Rutherford CL (1980): Localization of cyclic nucleotide phosphodiesterase in the multicellular stages of *Dictyostelium discoideum*. Differentiation 16:173–184.

Ceccarelli A, Mahbubani H, Williams JG (1991): Positively and negatively acting signals regulating stalk cell and anterior-like cell differentiation in *Dictyostelium*. Cell 65:983–989.

Coukell MB, Lappano S, Cameron AM (1983): Isolation and characterization of cAMP unresponsive (*frigid*) aggregation-deficient mutants of *Dictyostelium discoideum*. Dev Genet 3:283–297.

Datta S, Firtel RA (1987): Identification of the sequences controlling cyclic AMP regulation and cell-type specific expression of a prestalk-specific gene in *Dictyostelium discoideum*. Mol Cell Biol 7:149–159.

Devreotes P (1989): *Dictyostelium discoideum*: A model system for cell-cell interactions in development. Science 245:1054–1058.

Dharmawardhane S, Warren P, Hall A, Condeelis J (1989): Changes in the association of actin-binding proteins with the actin cytoskeleton during chemotactic stimulation of *Dictyostelium discoideum* amoebae. Cell Motil Cytoskel 13:57–63.

Early AE, Williams JG (1988): A *Dictyostelium* prespore specific gene is transcriptionally repressed by DIF in vitro. Development 103:519–524.

Esch RK, Firtel RA (1991): CAMP and cell sorting control the spatial expression of a developmentally essential cell-type-specific *ras* gene in *Dictyostelium*. Genes Dev 5:9–21.

Feit IN, Bonner JT, Suthers HB (1990): Regulation of the anterior-like cell state by ammonia in Dictyostelium discoideum. Dev Genet 11:442–446.

Firtel RA, Chapman AL (1990): A role for cAMP-dependent protein kinase in early *Dictyostelium* development. Genes Dev 4:18–28.

Firtel RA, van Haastert PJM, Kimmel AR, Devreotes P (1989): G-protein linked signal transduction pathways in development: *Dictyostelium* as an experimental system. Cell 58:235–239.

Fischer EH, Charbonneau H, Tonks NK (1991): Protein tyrosine phosphatases: a diverse family of intracellular and transmembrane enzymes. Science 253:401–406.

Fosnaugh KL, Loomis WF (1991): Coordinate regulation of the spore coat genes in *Dictyostelium discoideum*. Dev Gene 12:123–32.

Franke J, Faure M, Wu L, Hall AL, Podgorski GJ, Kessin RH (1991): Cyclic nucleotide phosphodiesterase of *Dictyostelium discoideum* and its glycoprotein inhibitor: Structure and expression of their genes. Dev Gene 12:104–12.

Gerisch G, Fromm H, Huesgen A, Wick U (1975): Control of cell-contact sites by cyclic AMP pulses in differentiating *Dictyostelium cells*. Nature 255:547–549.

Gomer RH, Datta S, Firtel RA (1986): Cellular and subcellular distribution of a cAMP-regulated prestalk protein and prespore protein in *Dictyostelium discoideum*. J Cell Biol 103:1999–2015.

Gomer RH, Firtel RA (1987): Cell-autonomous determination of cell-type choice in *Dictyostelium* development by cell-cycle phase. Science 237:758–762.

Gomer RH, Yuen IS, Firtel RA (1991): A secreted 80 kDa Mr protein mediates sensing of cell density and the onset of development in *Dictyostelium*. Development 112 (in press).

Green JBA, Smith JC (1990): Graded changes in dose of a *Xenopus* activin A homologue elicit stepwise transitions in embryonic cell fate. Nature 347:391–394

Gupta SK, Lowndes JM, Osawa S, Johnson GL (1991): Identification of G protein mutants encoding constitutively active and dominant negative a subunit peptides. In Mond JL, Cambier JC, Weiss (eds): "Advances in Regulation of Cell Growth." New York: Raven Press, pp 119–140.

Haberstroh L, Firtel RA (1990): A spatial gradient of expression of a cAMP-regulated prespore cell-type-specific gene in *Dictyostelium*. Genes Dev. 4:596–612.

Haberstroh L, Galindo J, Firtel RA (1991): Developmental and spatial regulation of a *Dictyostelium* prespore gene: *Cis*-acting elements and a cAMP-induced, developmentally regulated DNA-binding activity. Development (in press).

Hadwiger JA, Wilkie TM, Stratmann M, Firtel RA (1991): Identification of *Dictyostelium* Gα genes expressed during multicellular development. Proc Natl Acad Sci USA 88:8213–8217.

Hadwiger JA, Firtel RA (1992): Analysis of Gα4, a G-protein subunit required for multicellular development in *Dictyostelium*. Genes Dev 6:38–49.

Hall AL, Warren V, Condeelis J (1989): Transduction of the chemotactic signal to the actin cytoskeleton of *Dictyostelium discoideum*. Dev Biol 136:517–525.

Hanks SK, Quinn AM (1991): Protein kinase catalytic domain sequence database: Identification of conserved features of primary structure and classification of family members. Methods Enzymol 200:38-62.

Haribabu B, Dottin RP (1991): Identification of a protein kinase multigene family of *Dictyostelium discoideum*: Molecular cloning and expression of a cDNA encoding a developmentally regulated protein kinase. Proc Natl Acad Sci USA 88:1115–1119.

Hjorth AL, Khanna NC, Firtel RA (1989): A *trans*-acting factor required for cAMP-induced gene expression in *Dictyostelium* is regulated developmentally and induced by cAMP. Genes Dev 3:747–759.

Hjorth AL, Pears C, Williams JG, Firtel RA (1990): A developmentally regulated *trans*-acting factor recognizes dissimilar G/C rich elements controlling a class of cAMP-inducible *Dictyostelium* genes. Genes Dev 4:419–432.

Hunter T (1987): A thousand and one protein kinases. Cell 50:823–829.

Insall R, Kay RR (1990): A specific DIF binding protein in *Dictyostelium*. EMBO J 9:3323–3328.

Janssens PMW, van Haastert PJM (1987): Molecular basis of transmembrane signal transduction in *Dictyostelium discoideum*. Microbiol Rev. 51:396–418.

Jermyn KA, Berks M, Kay RR, Williams JG (1987): Two distinct classes of prestalk-enriched mRNA sequences in *Dictyostelium discoideum*. Development 100:745–755.

Jermyn KA, Williams JG (1991): An analysis of culmination in *Dictyostelium* using prestalk and stalk-specific cell autonomous markers. Development 111:779–787.

Kesbeke F, Snaar-Jagalska BE, van Haastert PJM (1988): Signal transduction in *Dictyostelium fgd* A mutants with a defective interaction between surface cAMP receptor and a GTP-binding regulatory protein. J Cell Biol 197:521–528.

Kimmel AR, Carlislile B (1986): A gene expressed in undifferentiated vegetative *Dictyostelium* is repressed by developmental pulses of cAMP and reinduced during dedifferentiation. Proc Natl Acad Sci USA 83:2506–2510.

Kimmel AR, Firtel RA (1991): Signal transduction pathways regulating development of *Dictyostelium discoideum*. Current Opinion Gene Dev 1 (in press).

Klein PS, Sun TJ, Saxe CL, Kimmel AR, Johnson RL, Deverotes PN (1988): A chemoattractant receptor controls development in *Dictyostelium discoideum*. Science 241:1467–1472.

Krefft M, Voet L, Gregg JH, Mairhofer H, Williams KL (1984): Evidence that positional information is used to establish the prestalk-prespore pattern in *Dictyostelium discoideum* aggregates. EMBO J 3:201–206.

Kumagai A, Hadwiger J, Pupillo M, Firtel RA (1991): Molecular analysis of two Ga protein subunits in *Dictyostelium* J Biol Chem 266:1220–1228.

Kumagai A, Pupillo M, Gundersen R, Miake-Lye R, Devreotes PN, Firtel RA (1989): Regulation and function of Ga protein subunits in *Dictyostelium*. Cell 57:265–275.

Leach CK, Ashworth JM, Garrod DR (1973): Cell sorting out during the differentiation of mixtures of metabolically distinct populations of *Dictyostelium discoideum*. J Ebryol Exp Morph 29:647–661.

Liu G, Newell PC (1988): Evidence that cyclic GMP regulates myosin interaction with the cytoskeleton during chemotaxis of *Dictyostelium*. J Cell Sci 90:123–129.

Loomis WF (ed) (1982): "The Development of *Dictyostelium Discoideum*." New York: Academic Press.

Majerfeld IH, Leichtling BH, Meligeni JA, Spitz E, Rickenberg HV (1984): A cytosolic cyclic AMP-dependent protein kinase in *Dictyostelium discoideum*. I. Properties. J Biol Chem 259:654–661.

Malkinson AM, Ashworth JM (1972): Extracellular concentrations of adenosine 3′5′ cyclic monophosphate during axenic growth of myxamoebae of the cellular slime mould *Dictyostlium discoideum*. Biochem J 133:601–603.

Mann SKO, Firtel RA (1987): Cyclic AMP regulation of early gene expression in *Dictyostelium discoideum*: Mediation via the cell surface cyclic AMP receptor. Mol Cell Biol 7:458–469.

Mann SKO, Firtel RA (1989): Two-phase regulatory pathway controls cAMP receptor-mediated expression of early genes in *Dictyostelium*. Proc Natl Acad Sci USA 86:1924–1928.

Mann SKO, Firtel RA (1991): A developmentally regulated putative serine/threonine protein kinase is essential for development in *Dictyostelium*. Mech Dev (in press).

Mann SKO, Pinko C, Firtel RA (1988): cAMP regulation of early gene expression in signal transduction mutants of *Dictyostelium*. Dev Biol 130:294–303.

May T. Blusch J, Sachse A, Nellen W (1991): A *cis*-acting element responsible for early gene induction by extracellular cAMP in *Dictyostelium discoideum*. Mech Dev 33:147–156.

Mee JD, Tortolo C, Coukell MB (1986): Chemotaxis associated properties of separated prestalk and prespore cells. Biochem Cell Biol 64:722–732.

Mehdy MC, Firtel RA (1985): A secreted factor and cyclic AMP jointly regulate cell-type-specific gene expression in *Dictyostelium discoideum*. Mol Cell Biol 5:705–713.

Mehdy MC, Ratner D, Firtel RA (1983): Induction and modulation of cell-type-specific gene expression in *Dictyostelium*. Cell 32:761–771.

Morris HR, Taylor GW, Masento MS, Jermyn KA, Kay RR (1987): Chemical structure of the morphogen differentiation inducing factor from *Dictyostelium discoideum*. Nature 328:811–814.

Mutzel R, Lacombe M-L, Simon M-N, de Gunzburg J, Veron M (1987): Cloning and cDNA sequence of the regulatory subunit of cAMP-dependent protein kinase from *Dictyostelium discoideum*. Proc Natl Acad Sci USA 84:6–10.

Okaichi K, Cubitt AB, Pitt G, Firtel RA (1992): Amino acid substitutions in the *Dictyostelium* Gα subunit Gα2 produce dominant negative phenotypes and inhibit the activation of adenylyl cyclase, guanylyl cyclase, and phospholipase C. Mol Biol Cell 3 (in press).

Otte AP, Plomp MJE, Arents JC, Jansens PMW, van Driel R (1986): Production and turnover of cAMP signals by prestalk and prespore cells in *Dictyostelium discoideum* cell aggregates. Differentiation 32:185–191.

Pears CJ, Williams JG (1987): Identification of a DNA sequence element required for efficient expression of a developmentally regulated and cAMP-inducible gene of *Dictyostelium discoideum*. EMBO J 6:195–200.

Pears CJ, Williams JG (1988): Multiple copies of a G-rich element upstream of a cAMP-inducible *Dictyostelium* gene are necessary but not sufficient for efficient gene expression. Nucleic Acids Res 16:8467–8486.

Podgorski GJ, Faure M, Franke J, Kessin RH (1988): The cyclic nucleotide phosphodiesterase of *Dictyostelium discoideum*: The structure of the gene and its regulation and role in development. Dev Gene 9:267–78.

Pupillo M, Kumagai A, Pitt GS, Firtel RA, Devreotes PN (1989): Multiple α subunits of guanine nucleotide binding proteins in *Dictyostelium*. Proc Natl Acad Sci USA 86:4892–4896.

Raper KB (1940): Pseudoplasmodium formation and organization in *Dictyostelium discoideum*. J Elisha Mitchell Sci Soc 56:241–282.

Reymond CD, Gomer RH, Mehdy M, Firtel RA (1984): Developmental regulation of a *Dictyostelium* gene encoding a protein homologous to mammalian *ras* protein. Cell 39:141–148.

Reymond CD, Gomer RH, Nellen W, Theibert A, Devreotes P, Firtel RA (1986): Phenotypic changes induced by a mutated *ras* gene during the development of *Dictyostelium* transformants. Nature 323:340–343.

Riley BB, Barclay SL (1990): Ammonia promotes accumulation of intracellular cAMP in differentiating amoebae of *Dictyostelium discoideum*. Development 109:715–22.

Robbins SM, Williams JG, Jermyn KA, Spiegelman GB, Weeks G (1989): Growing and developing *Dictyostelium* cells express different *ras* genes. Proc Natl Acad Sci USA 86:938–942.

Rubin J, Robertson A (1975): The tip of the *Dictyostelium discoideum* pseudoplasmodium as an organizer. J Embryol Exp Morph 36:663–668.

Rubino S, Mann SK, Hori RT, Pinko C, Firtel RA (1989): Molecular analysis of a developmentally regulated gene required for *Dictyostelium* aggregation. Dev Biol 131:27–36.
Saxe CL III, Johnson RL, Devreotes PN, Kimmel AR (1991a): Expression of a cAMP receptor gene of *Dictyostelium* and evidence for a multigene family. Genes Dev 5:1–8.
Saxe CL III, Johnson RL, Devreotes PN, Kimmel AR (1991b): Multiple genes for cell surface cAMP receptors in *Dictyostelium discoideum*. Dev Genet 12:6–13.
Schaap P (1983): Quantitative analysis of the spatial distribution of ultrastructural differentiation markers during development in *Dictyostelium discoideum*. Roux's Arch Dev Biol 192:86–94.
Schaap P, Wang M (1986): Interactions between adenosine and oscillatory cAMP signalling regulate size and pattern in *Dictyostelium discoideum*. Cell 45:137–144.
Schaller HC, Hoffmeister SAH, Dubel S (1989): Role of the neuropeptide head activator for growth and development in hydra and mammals. Development 107 (Suppl):99–107.
Schindler J, Sussman M (1977): Ammonia determines the choice of morphogenetic pathways in *Dictyostelium discoideum*. J Molec Biol 116:161–170.
Schweiger A, Mihalache O, Muhr A, Adrian I (1990): Phosphotyrosine-containing proteins in *Dictyostelium discoideum*. FEBS Lett 268:199–202.
Simon MI, Strathmann MP, Gautam N (1991): Diversity of G proteins in signal transduction. Science 252:802–8.
Simon MN, Driscoll D, Mutzel R, Part D, Williams J, Veron M (1989): Overproduction of the regulatory subunit of the cAMP dependent protein kinase blocks the differentiation of *Dictyostelium discoideum*. EMBO J 8:2039–2043.
Siu CH (1990): Cell-cell adhesion molecules in *Dictyostelium*. Bioessays 12:357–362.
Snaar-Jagalska BE, van Haastert PJM (1988): G-proteins in the signal-transduction pathway of *Dictyostelium discoideum*. Dev Genet 9:215–226.
Sternfeld J, David C (1982): Fate and regulation of anterior-like cells in *Dictyostelium* slugs. Dev Biol 93:111–118.
Summerbell D, Maden M (1990): Retinoic acid, a developmental signalling molecule. Trends Neurosci 13:142–147.
Sun TJ, Devreotes PN. (1991): Gene targeting of the aggregation stage cAMP receptor cAR1 in *Dictyostelium*. Genes Dev 5:572–82.
Takeichi M (1990): Cadherins: A molecular family important in selective cell-cell adhesion. Annu Rev Biochem 59:237–521.
Takeuchi I (1963): Immunochemical and immunohistochemical studies on the development of the cellular slime mold *Dictyostelium discoideum*. Dev Biol 8:1–26.
Tan JL, Spudich JA (1990): Developmentally regulated protein-tyrosine kinase genes in *Dictyostelium discoideum*. Mol Cell Biol 10:3578–3583.
Theibert A, Devreotes PN (1984): Adenosine and its derivatives inhibit the cAMP signaling response in *Dictyostelium discoideum*. Dev Biol 106:166–173.
Theibert A, Devreotes PN (1986): Surface receptor-mediated activation of adenylate cyclase in *Dictyostelium*: regulation by guanine nucleotides in wild-type cells and aggregation deficient mutants. J Biol Chem 261:15121–15125.
Town CD, Gross JD, Kay RR (1976): Cell differentiation without morphogenesis in *Dictyostelium discoideum*. Nature 262:717–719.
Vaughan R, Devreotes PN (1988): Ligand-induced phosphorylation of the cAMP receptor from *Dictyostelium discoideum*. J Biol Chem 263:14538–14543.
Weijer CJ, Duschl G, David CN (1984): Dependence of cell-type proportioning and sorting on cell cycle phase in *Dictyostelium discoideum*. J Cell Sci 70:133–145.
Williams JG (1988): The role of diffusible molecules in regulating the cellular differentiation of *Dictyostelium discoideum*. Development 103:1–16.
Williams JG, Ceccarelli A, McRobbie S, Mahbubani H, Kay RR, Early A, Berks M, Jermyn KA (1987): Direct induction of *Dictyostelium* prestalk gene expression by DIF provides evidence that DIF is a morphogen. Cell 49:185–192.

Williams JG, Duffy KT, Lane DP, McRobbie SJ, Harwood AJ, Traynor D, Kay RR, Jermyn KA (1989): Origins of the prestalk-prespore pattern in *Dictyostelium* development. Cell 59:1157–1163.

Williams KL, Vardy PH, Segel LA (1986): Cell migrations during morphogenesis: Some clues from the slug of *Dictyostelium discoideum*. Bioessays 5:148–152.

Evolutionary Conservation of Developmental Mechanisms, pages 185–198

12. Developmental and Molecular Responses to Touch in Plants

Janet Braam

Department of Biochemistry and Cell Biology, Rice University, Houston, Texas 77251

IMPACT OF THE ENVIRONMENT ON PLANT DEVELOPMENT

Despite their passive appearance, plants sense and actively respond to many environmental stimuli. Plants, unlike many animals, cannot seek shelter from environmental stresses; instead they often adapt to the changes and stresses of their environment to survive. One way that plants adapt to the environment is through developmental alterations. That is, plant development is usually plastic because it can be dramatically altered as a result of the environmental conditions. Plasticity of development is thought to be particularly important for sessile organisms because it allows adaptation to inescapable environmental stresses. Clearly, understanding how plants respond to the environment is essential to the study of plant growth and development.

MORPHOLOGICAL AND PHYSIOLOGICAL RESPONSES OF PLANTS TO MECHANICAL STIMULI

Mechanical stimuli such as rain, wind, changes in osmotic conditions, obstructions in the soil, and gravity are ubiquitous and are known to result in morphological alterations in plants (Darwin, 1881; Morgan, 1984; Biddington, 1986). In addition, plants sense and respond to the simple mechanical stimulation of touch. Darwin (1881) was one of the first to document the touch sensitivity of roots. When small pieces of cardboard were attached to the sides of pea root tips, the growth of the roots was altered resulting in bending and curling of the root away from the point of contact. Another fascinating touch response is demonstrated by Venus's flytrap (*Dionaea muscipula*); the specialized bilobed leaves rapidly snap shut to catch prey (Sibaoka, 1969). The sensitive plant (*Mimosa pudica*) also displays a very rapid touch response; the specialized pulvini at the base of the leaflets allow rapid leaflet closure in response to a touch stimulus (Sibaoka, 1969). In addition, there are examples of pea tendrils and stamens of flowers that bend in response to a touch stimulation. These touch responses enable the tendrils to attach to nearby supports and the stamens to deposit pollen on visiting insects.

More surprisingly, 80% of the "nonspecialized" plant species tested (Jaffe, 1973) also respond to touch, sometimes quite dramatically. The touch-induced response is morphological, generally including a decrease in longitudinal growth and an increase in radial expansion (Jaffe, 1973). These developmental alterations have been termed thigmomorphogenesis (Jaffe, 1973) and have been extensively documented in a variety of plants (Jaffe, 1973). Similar growth responses, called seismomorphogenesis (Mitchell et al., 1975), are found in plants exposed to vibrations or shaking. An example of thigmomorphogenic growth is illustrated in Figure 1. The *Arabidopsis* plant on the right was touched two times a day throughout its life, the plant on the right is the untreated control (Braam and Davis, 1990a). The touch stimulus consisted of gently moving the rosette leaves back and forth to cause bending of the stems and petioles. Thigmomorphogenesis is likely an adaptive response, enabling greater resistance to environmental stimuli like wind. For

Fig. 1. *Arabidopsis* displays developmental alterations as a result of touch stimulation. *Arabidopsis* plants at 6 weeks of age. The plant on the left was touched twice daily; plant on the right is the untreated control. (Reprinted from Braam and Davis, 1990a, with permission from Cell Press.)

example, trees allowed to sway in the wind develop shorter, thicker trunks than trees prevented from swaying (Jacobs, 1954). The shorter, stockier trees are thus less likely to be damaged by the wind.

There are several very rapid physiological responses to mechanical stimulation. Jaffe (1976a) showed that within 3 minutes of touch stimulation of beans (*Phaseolus vulgaris*), there is a brief increase in growth rate, followed by a complete inhibition of elongation for 15 to 30 minutes; normal growth resumes after 3 to 4 days. Touch stimulation also inhibits growth of portions of the plant, especially young tissues, that are not stimulated directly (Jaffe, 1976a). This observation has led to the speculation that a signal may be communicated to nonstressed areas. Other rapid responses in plants following touch stimulation include an immediate drop in electrical resistance in the stem (Jaffe, 1976b), an increase in small voltage transients (Pickard, 1971), callose synthesis (Jaffe et al., 1985), localized blockage of phloem transport (Jaeger et al., 1988), and within an hour, an increase in ethylene production (Biro and Jaffe, 1984). What intracellular signals are generated following mechanical stimulation that lead to these varied responses?

CALCIUM INVOLVEMENT IN PLANT RESPONSES TO MECHANICAL STIMULI

The signal transduction pathway by which plants sense and respond to mechanical stimuli is not known; however, evidence has accumulated that suggests Ca^{2+} may act as second messenger. First, detection of Ca^{2+} levels in touch-stimulated tobacco seedlings indicate that touch stimulation results in rapid and transient fluctuations in Ca^{2+} levels (Knight et al., 1991). In addition, cytoplasmic Ca^{2+} levels increase following changes in orientation of plants to the gravitational field (Gehring et al., 1990), or changes in osmotic pressure (Okazaki et al., 1987). Application of calmodulin inhibitors or EGTA, a Ca^{2+}-specific chelator, inhibits thigmomorphogenesis (Jones and Mitchell, 1989) and gravitropism (Evans et al., 1986), whereas Ca^{2+} activates the responses (Jones and Mitchell, 1989; Evans et al., 1986). Lowering Ca^{2+} with EGTA also prevents turgor regulation (Okazaki and Tazawa 1986).

Further, Ca^{2+} changes have been implicated in eliciting each of the processes known to occur in response to touch stimulation, including growth inhibition (Roux and Slocum, 1982), ion flux (Pickard, 1984), callose production (Kauss, 1987), and ethylene production (Pickard, 1984).

The demonstration that a variety of mechanical stimuli induce the expression of three genes encoding calmodulin-related proteins (*TCH 1, 2,* and *3*) in *Arabidopsis* (Braam and Davis, 1990a) also suggests a role for Ca^{2+} in the mechanosensory signal transduction pathway. This molecular response to touch stimulation and its significance is reviewed below.

MOLECULAR RESPONSES TO MECHANICAL STIMULI IN *ARABIDOPSIS THALIANA*

A rapid and strong induction of gene expression following touch stimulation of *Arabidopsis* plants was discovered using standard differential cDNA screening (Braam and Davis, 1990a). Five cDNAs were obtained that represented different mRNAs that are found in greater abundance in touch-stimulated plants. The genes that encode these mRNAs have been termed the touch (*TCH*) genes (Braam and Davis, 1990a). Figure 2 shows that *TCH* mRNAs increase in abundance in plants subjected to a variety of stimuli (Braam and Davis, 1990a). Individual pots of plants were subjected to the indicated treatments, collected 30 minutes later, and poly $(A)^+$ RNA levels were analyzed by northern blots. Levels of expression of the *TCH 1, 2, 3,* and *4* genes are increased after spraying plants with water, subirrigation, wounding, touching, and exposure to darkness. The *TCH* genes are also regulated

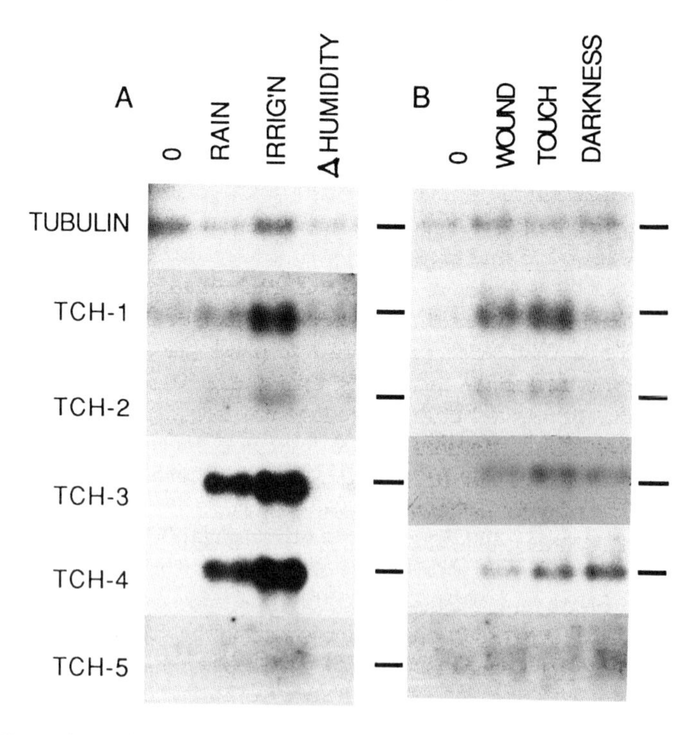

Fig. 2. Expression of the *TCH* genes is induced by a variety of stimuli. Composite northern blots of poly $(A)^+$ RNA purified from untreated plants (A,B: lanes "O") and from plants harvested 30 min after the indicated treatments. Treatments are indicated above each lane and are described in the text and, more fully, in Braam and Davis (1990a). The RNA was size separated on formaldehyde gels, the gel was blotted to a filter, and the filter was hybridized sequentially with each of the probes indicated at the left. Band intensities in different panels reflect different lengths of exposure to film. (Reprinted from Braam and Davis, 1990a, with permission from Cell Press.)

following wind treatment, but are not affected when plants are carefully moved from one location to a similar location (Braam and Davis, 1990a). Thus, the *TCH* genes are induced by touch and other mechanical perturbations, but also by seemingly unrelated stimuli, such as darkness. Currently, research is underway to determine whether these unrelated stimuli regulate *TCH* gene induction through similar or diverse mechanisms. In addition, we are investigating whether the *TCH* mRNAs accumulate as a result of an increase in rate of transcription initiation or modulation of mRNA stability.

All of the *TCH* mRNAs begin to accumulate within 10 minutes after touch stimulation; however, maximum levels of mRNA accumulation are found at different times for the individual *TCH* genes (Braam and Davis, 1990a) (see Fig. 3). *TCH 1* and *TCH 4* show maximal mRNA levels 10 minutes after stimulation, whereas *TCH 3* and *TCH 5* mRNAs accumulate to their highest levels after 30 minutes, and *TCH 2* mRNA levels are maximal 30 minutes to 1 hour after stimulation. The nonconformity of the kinetics of *TCH* gene response suggests the possibility that variations exist in the mechanisms of regulation of the different *TCH* genes. The return of *TCH* transcripts to basal levels 1 to 2 hours after stimulation indicates that the *TCH* mRNAs are relatively unstable.

Using the cloned *TCH* genes as tools, the response to mechanical stimulation can be further characterized. For example, to determine whether the

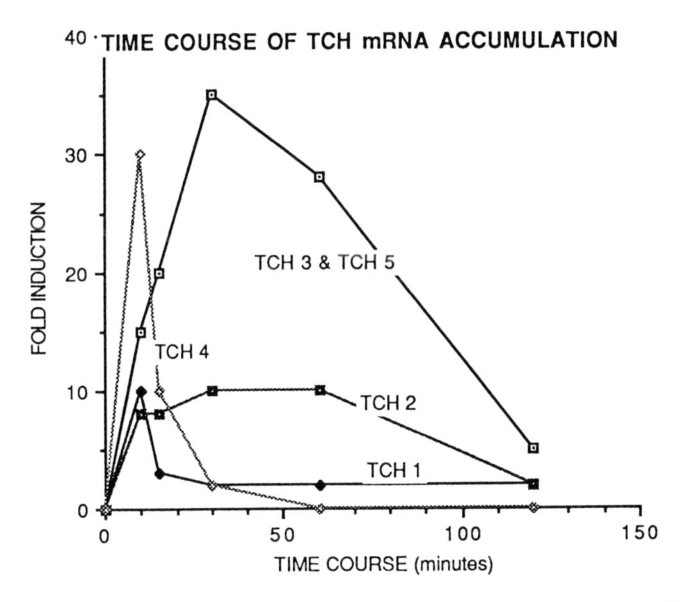

Fig. 3. Kinetics of induction of the *TCH* genes following touch stimulation. Poly $(A)^+$ RNA was purified from untreated plants, and from plants harvested at the indicated times following touch treatment. The hybridization bands from northern blot analysis were quantitated by densitometry and graphed. (Adapted from Braam and Davis, 1990a, with permission from Cell Press.)

mechanosensory system in plants is capable of detecting differences in the strength of the stimulation, *TCH* mRNA levels were compared in plants that were subjected to increasing doses of touch. Increased intensity of touch stimulation results in relative increases of *TCH* mRNA levels (Braam and Davis, 1990a) (Fig.4). Therefore, a characteristic expected of the machinery that perceives touch is the ability to detect the strength of the stimulation and to transmit quantitatively variable signals. Further, the regulatory mechanism governing *TCH* gene expression is not an on-off switch, but must act proportionally to a quantitative signal.

Figure 5 compares the response following a single stimulation to that following 30 minutes of repetitive stimulation; generally, repetitive stimulation results in a response that is stronger and more prolonged (Braam and Davis, 1990b). Plants exposed to touch stimulation at time zero and again at either 2 hours or 3 hours show *TCH* mRNA accumulation 30 minutes after each stimulation (Braam and Davis, 1990b) (Fig.6). The fact that the second response, however, is generally weaker than the initial response shows that there is partial desensitization or habituation.

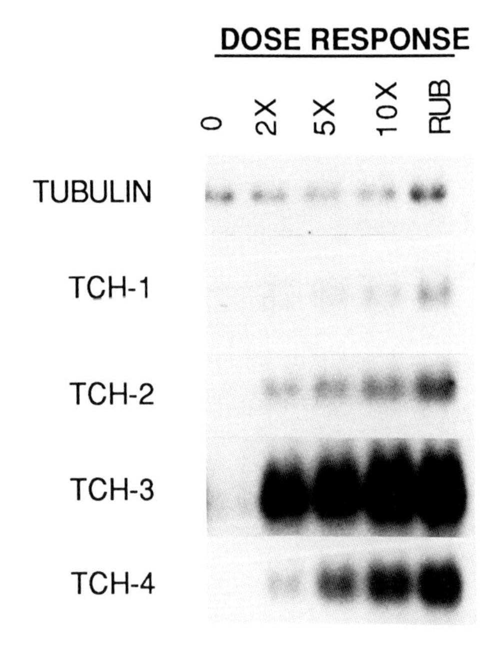

Fig. 4. The accumulation levels of *TCH* transcripts are proportional to the strength of the stimulation. Composite northern blots of RNA purified from untreated plants (lane 0) and from plants that were stimulated by: touching the rosette leaves so as to bend the petioles and stems back and forth twice (2X), five times (5X), or ten times (10X); or gently rubbing individual leaves (RUB). The RNA was size separated by gel electrophoresis, blotted to a filter and hybridized sequentially with the cDNA probes listed at the left. (Reprinted from Braam and Davis, 1990a, with permission from Cell Press).

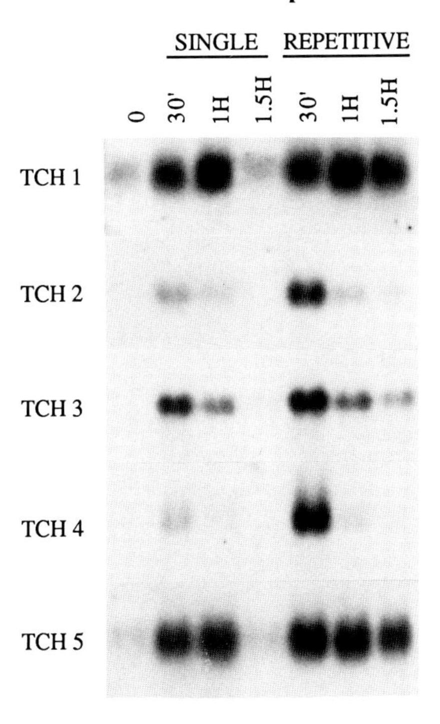

Fig. 5. Repetitive stimulation results in a stronger and more prolonged response. Northern blot analysis of poly $(A)^+$ RNA purified from untreated plants (lane 0), or from plants harvested at the indicated times (30', 1H or 1.5H), following a single touch treatment (SINGLE), or following seven touch treatments, each at 5-min intervals over 30 min (REPETITIVE). The filters were probed sequentially with the cDNAs indicated at the left. (Reprinted from Braam and Davis, 1990b.)

The sequences of three of the *TCH* genes encode proteins closely related to calmodulin suggesting that these gene products may play central roles in signal transduction. The deduced amino acid sequences of *TCH 1, 2,* and *3* share high degrees of similarities to wheat calmodulin (Braam and Davis, 1990a). The *TCH 1* gene most likely encodes an *Aradidopsis* calmodulin, and *TCH 2* and *TCH 3* encode proteins closely related to calmodulin (Braam and Davis, 1990a). Recently, another gene with a 13% amino acid divergence from *TCH 1* has been isolated from *Arabidopsis* and shown to be similarly induced in expression by touch stimulation (Ling et al., 1991).

FUNCTIONS OF CALMODULIN AND CALMODULIN-RELATED PROTEINS

The identities of the *TCH* genes and the strong and rapid induction of their expression levels suggest that, following mechanical stimulation, there is an immediate but temporary requirement for increased synthesis of Ca^{2+}-bind-

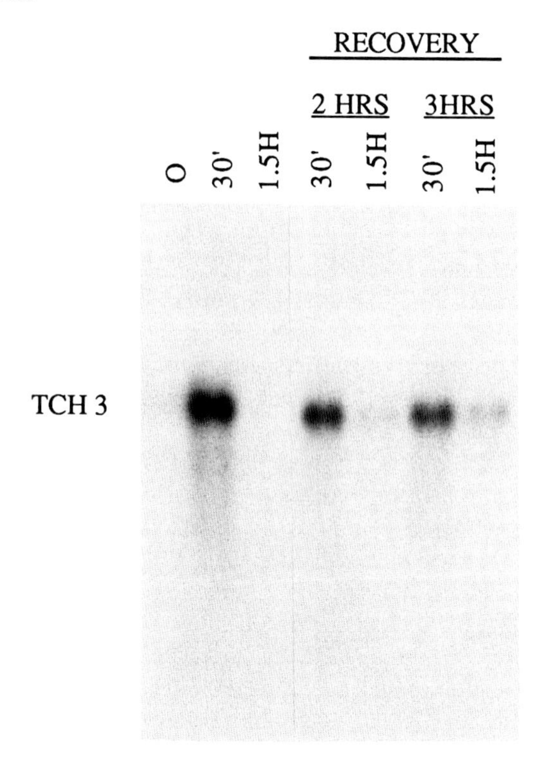

Fig. 6. Desensitization of touch-induction of *TCH3* gene expression. Northern blot analysis of poly $(A)^+$ RNA purified from untreated plants (lane 0), or from plants harvested at the indicated times (30' or 1.5H) after touch treatment at time 0 alone (lanes 30' and 1.5H), at time 0 and again 2 hours later (2 HRS, 30' and 1.5H), or at time 0 and again 3 hours later (3 HRS, 30' and 1.5H). (Reprinted from Braam and Davis, 1990b.)

ing proteins. The functions of the TCH proteins and their roles in the physiological and morphological responses of *Arabidopsis* to environmental stimuli are not known. However, TCH protein functions may be related to those of calmodulin and calmodulin-related proteins of other organisms.

The sequence conservation and ubiquity of calmodulin among eukaryotes suggest that it is involved in fundamental cellular processes (reviewed in Cheung, 1980; Stoclet et al., 1987; Cohen and Klee, 1988; Means, 1988). Calmodulin is thought to be the primary intracellular Ca^{2+} receptor and to function in the mediation of cellular responses to fluctuations in free cytoplasmic Ca^{2+}. Calmodulin may also perform essential functions that are independent of Ca^{2+} binding (Geiser et al., 1991). Ca^{2+} fluctuations function as a second messenger, translating external stimuli into intracellular signals (reviewed in Hepler and Wayne, 1985). In plants, Ca^{2+} may come from millimolar Ca^{2+} stores, such as the vacuole and the cell wall (Hepler and Wayne, 1985). Because Ca^{2+} is cytotoxic at high levels, the free Ca^{2+} in the cytoplasm is kept below micromolar levels. One reason that Ca^{2+} is an ideal second messenger is because this large concentration gradient allows a dramatic (10- to 100-fold) change in Ca^{2+} without an energy expense (Hepler and Wayne, 1985). The cell allows Ca^{2+} to flow down its concentration gradient into the cytoplasm as a signal that a certain stimulus has occurred. Ca^{2+}-binding proteins with affinities of 1 to 10 μM, such as calmodulin, probably detect Ca^{2+} increases and mediate cellular responses. Calmodulin is known to modulate the activity of at least 20 different enzymes in vitro, such as phosphodiesterases, adenylate cyclases, Ca^{2+}-ATPases, and protein kinases; many of the substrates are key regulators of a variety of cellular processes (reviewed in Cohen and Klee, 1988). The TCH 1 protein, being virtually identical to calmodulin (Braam and Davis, 1990a), may play a role in a signal transduction cascade and act as a Ca^{2+}-dependent regulator of target enzymes.

Detection of intracellular Ca^{2+} levels suggest that mechanical stimulation can lead to fluctuations in Ca^{2+}. Knight et al. (1991) generated transgenic tobacco plants that express the gene encoding the Ca^{2+}-sensitive luminescent protein, aequorin, to monitor Ca^{2+} changes in intact plants. These researchers found that touch stimulation leads to rapid, significant Ca^{2+} increases in the tobacco seedlings. Ca^{2+} fluctuations following mechanical stress are not unique to plant cells. Wirtz and Dobbs (1990) reported cytoplasmic Ca^{2+} increases as a result of single stretches of individual lung cells, and the increases in Ca^{2+} observed were proportional to the extent of stretching.

It is possible that increases in cytoplasmic Ca^{2+} may directly or indirectly lead to increased expression of genes encoding calmodulin-related proteins, thus ensuring that enough Ca^{2+}-binding proteins are produced to accommodate increases in Ca^{2+}. Recent experiments demonstrate that the *TCH* genes are up-regulated following stimuli that evoke [Ca^{2+}] changes (Braam, 1992). This regulatory circuit would be crucial to maintaining the effectiveness of Ca^{2+} as a transient cellular signal and to ensuring that Ca^{2+} does not reach toxic levels.

Both calmodulin and calmodulin-related proteins have been implicated in regulating microtubule organization. Calmodulin enhances the Ca^{2+}-induced dissociation of microtubules (Cheung, 1980; Cohen and Klee, 1988). The calmodulin-related protein encoded by *CDC 31* of *Saccharomyces* is essential for the duplication of the spindle pole body during the cell cycle (Baum et al., 1986). A major protein in the microtubule organizing center of *Chlamydomonas* is a calmodulin-related protein called centrin or caltractin (reviewed in Lee and Huang, 1990). This protein is thought to form a Ca^{2+}-sensitive contractile fiber that may determine the orientation of cell division. Calmodulin in plant cells has been immunolocalized to the microtubule organizing centers (for example, see Vantard et al., 1985; Wick et al., 1985). The idea that the *TCH* gene products may be involved in microtubule organization is attractive because the arrangement of microtubules in plant cells determines the axis of cell expansion (Sinnot, 1960).

Thigmomorphogenesis generally results in a change from an elongation growth to radial growth (Jaffe, 1973), so a change in microtubule organization would probably be required. Thus, thigmomorphogenesis would require at least a partial disassembly of microtubules.

Calmodulin has been shown to be essential for the touch response of *Paramecium*. Kung (1971a, 1971b) identified mutants of *Paramecium* that show altered behavior following a mechanical stimulation. When *Paramecia* encounter an obstacle when swimming, ciliary action is transiently reversed, usually resulting in a change in direction of swimming. This behavior is correlated with an action potential resulting from the successive gating of ion channels (Saimi and Kung, 1980). Two types of mutants have been identified: underresponders that fail to alter the direction of swimming upon hitting an obstacle, and overreactive mutants that are delayed in the recovery after encountering mechanical stimulation. These phenotypes are correlated with defective Ca^{2+}-dependent Na channels and Ca^{2+}-dependent K channels, respectively. The channel failure, however, is probably due to defective regulation by calmodulin because the mutations are found in the calmodulin gene (Hinrichsen et al., 1986; Schaefer et al., 1987; Lukas et al., 1989; Kink et al., 1990). Therefore, calmodulin appears to be involved in the touch responses of two quite unrelated organisms, *Paramecium* and *Arabidopsis*. If calmodulin functions have been conserved in the touch response between plants and *Paramecium*, the calmodulin-related TCH proteins of *Arabidopsis* may have a role in regulating ion channels.

Calmodulin-related proteins have been identified in a large number of organisms (reviewed in Heizmann and Hunziker, 1991). The proposed functions of these proteins include sequestering, buffering, or transporting Ca^{2+}. Similarly, the calmodulin-related TCH proteins may bind Ca^{2+} to maintain Ca^{2+} homeostasis and allow cells to return quickly to the physiological resting state. Cells possess an extraordinary capacity to recover from stimulation and adapt to steady stimulation. This is critical in order for cells to remain sensitive to changes in the environment even against a background level of constant stimulation.

As yet, there is no evidence that the products of the TCH genes are involved in eliciting the thigmomorphogenic responses of *Arabidopsis*. However, Jones and Mitchell (1989) have observed a partial inhibition of the thigmomorphogenic response in soybeans as a result of treatment with calmodulin inhibitors, suggesting that calmodulin function is at least partially required for touch-induced growth inhibition in plants.

PERCEPTION OF MECHANICAL STIMULATION

It has been proposed that plants sense touch, wind, osmotic pressure, and gravity through a common mechanism based on the mechanical properties of the stimuli (Edwards and Pickard, 1987). The mechanism of perception of mechanical stimulation in biological systems is uncertain. However, mechanical distortion of cells is known to result in immediate inward electrical current (for example, see Julian and Goldman, 1962; Benolken and Jacobson, 1970), and the discovery of a plasma membrane ion channel that opens specifically in response to membrane stretching (Guharay and Sachs, 1984) provides an attractive candidate for the mechanosensor. These stretch-activated (SA) channels are generally not ion selective (Sachs, 1988), and may provide a pathway for significant Ca^{2+} importation into the cytoplasm (Yang and Sachs, 1989). The probability of SA channel opening is proportional to an exponential function of the stimulus strength (Gustin et al., 1988); however, SA channels adapt (or become desensitized) to a sustained stimulus by exhibiting a decrease in pressure sensitivity (Gustin et al., 1988). Many animal cell types, including those of skeletal muscle, nerve, epithelium, and heart, have SA channels (reviewed in Sachs, 1988; Saimi et al., 1988). In addition, SA channels have been found in higher plants (Edwards and Pickard, 1987), yeast (Gustin et al., 1988), and *E. coli* (Martinac et al., 1987). Because of this phylogenic conservation, it has been proposed that SA channels may have been essential for maintaining osmotic conditions in primitive cells and, in evolutionarily more advanced organisms, may also function in sensing sound, vibration, touch, gravity, and balance (Sachs, 1988; Saimi et al., 1988).

THE PROBLEM OF SPECIFICITY IN SIGNAL TRANSDUCTION BY Ca^{+}

The data described above are consistent with a hypothesis that Ca^{2+} fluctuations may act as second messengers to mediate cellular responses, including *TCH* gene regulation, to mechanical environmental stimuli. However, this putative role would probably not be specific for the mechanosensory pathway because Ca^{2+} has also been implicated in plant responses to a number of nonmechanical stimuli, including light, temperature, and hormones (Hepler and Wayne, 1985). Similarly, Ca^{2+} is thought to function in animal cell responses to mechanical and nonmechanical stimuli. This ubiquitous use of Ca^{2+} as a second messenger introduces two important and yet unsolved questions: (1) how do so many diverse stimuli generate a common second messenger?; (2) how are the appropriate and specific long-term responses to different stimuli specified when signals are transduced with a common second messenger? Answers to these questions are needed to generate a comprehensive picture of the pivotal process of signal transduction, not only in plant but also in animal cells.

The discovery of the regulation of expression of the *Arabidopsis TCH* genes has uncovered one step that appears to be common in the chain of events that follows exposure of plants to several stimuli, that is, the rapid and strong induction of expression of the five *TCH* genes (Braam and Davis, 1990a). The inductive stimuli include mechanical and nonmechanical stimuli that appear to be unrelated except that Ca^{2+} has been implicated in mediating the respective, but again unrelated, cellular responses. With the *TCH* genes cloned, we have molecular tools necessary to characterize the preceding and subsequent steps in the sensory and response pathways for these inductive stimuli. Investigating the prior steps will enable us to compare the mechanisms of gene activation and the *cis* and *trans* regulatory elements utilized for different stimuli. In addition, we will determine whether Ca^{2+} changes are necessary and/or sufficient for each response. Differences in subsequent steps, such as mRNA and protein accumulation and localization, may help to elucidate how distinct long-term physiological responses are specified. In addition, the isolation of mutants defective in the mechanosensory pathway will be critical in the identification of the steps and components which function in perception and response to mechanical stimuli.

REFERENCES

Baum P, Furlong C, Byers B (1986): Yeast gene required for spindle pole body duplication: Homology of its product with Ca^{2+}-binding proteins. Proc Natl Acad Sci USA 83:5512–5516.

Benolken RM, Jacobson SL (1970): Response properties of a sensory hair excised from Venus's fly trap. J Gen Physiol 56:64–82.

Biddington NL (1986): The effects of mechanically-induced stress in plants—a review. Plant Growth Reg 4:103–123.

Biro RL, Jaffe MJ (1984): Thigmomorphogenesis: Ethylene evolution and its role in the changes observed in mechanically perturbed bean plants. Physiol Planta 62:289–296.

Braam J (1992): Regulated expression of the calmodulin-related *TCH* genes in *cultured Arabidopsis* cells: Induction by calcium and heat shock. Proc Natl Acad Sci USA 89:3213–3216.

Braam J, Davis RW (1990a): Rain-, wind- and touch-induced expression of calmodulin and calmodulin-related genes in *Arabidopsis*. Cell 60:357–364.

Braam J, Davis RW (1990b): The mechanosensory pathway in *Arabidopsis*: Touch-induced regulation of expression of calmodulin and calmodulin-related genes and alterations of development. In Randall DD, Blevins DG, (eds): "Current Topics in Plant Biochemistry and Physiology, Vol 9." Columbia: University of Missouri Press, pp. 85–100.

Cheung WY (ed) (1980): "Calmodulin. Calcium and Cell Function, Vol 1." New York: Academic Press.

Cohen P, Klee CB (eds) (1988): "Calmodulin, Molecular Aspects of Cellular Regulation." New York: Elsevier.

Darwin C (1881): "The Power of Movement in Plants." New York: Appleton.

Edwards KL, Pickard BG (1987): Detection and transduction of physical stimuli in plants. In Greppen H, Milet B, Wagner E (eds): "The Cell Surface in Signal Transduction." Berlin: Springer-Verlag, pp 41–66.

Evans ML, Moore R, Hasenstein K-H (1986): How roots respond to gravity. Sci Amer 255:112–119.

Gehring CA, Williams DA, Cody SH, Parish RW (1990): Phototropism and geotropism in maize coleoptiles are spatially correlated with increases in cytosolic free calcium. Nature 345:528–530.

Geiser JR, Vantuinen D, Brockerhoff SE, Neff MM, Davis TN (1991): Can calmodulin function without binding calcium? Cell 65:949–959.

Guharay F, Sachs F (1984): Stretch-activated single ion channels currents in tissue-cultured embryonic chick skeletal muscle. J Physiol (London) 352:685–701.
Gustin MC, Zhou X-L, Martinac B, Kung C (1988): A mechanosensitive ion channel in the yeast plasma membrane. Science 242:762–765.
Heizmann CW, Hunziker W (1991): Intracellular calcium-binding proteins: More sites than insights. Trends Biochem Sci 16:98–103.
Hepler PK, Wayne RO (1985): Calcium and plant development. Ann Rev Plant Physiol 36:397–439.
Hinrichsen RD, Burgess-Cassler A, Soltvedt BC, Hennessey R, Kung C (1986): Restoration by calmodulin of a Ca^{2+}-dependent K^+ current missing in a mutant of *Paramecium*. Science 232:503–506.
Huang B, Mengersen A, Lee VD (1988): Molecular cloning of cDNA for caltractin, a basal body-associated Ca^{2+}-binding protein: Homology in its protein sequence with calmodulin and the yeast *CDC 31* gene product. J Cell Biol 107:133–140.
Jacobs MR (1954): The effect of wind sway on the form and development of *Pinus radiata*. Austral J Bot 2:35–51.
Jaeger CH, Goeschl JD, Magnuson CE, Fares Y, Strain BR (1988): Short-term responses of phloem transport to mechanical perturbation. Physiol Planta 72:588–594.
Jaffe MJ (1973): Thigmomorphogenesis: The response of plant growth and development to mechanical stimulation. Planta 114:143–157.
Jaffe MJ (1976a): Thigmomorphogenesis: A detailed characterization of the response of beans to mechanical stimulation. Z Pflanzenphysiol 77:437–453.
Jaffe MJ (1976b): Thigmomorphogenesis: Electrical resistance and mechanical correlates of the early events of growth retardation due to mechanical stimulation in beans. Z Pflanzenphysiol 78:24–32.
Jaffe MJ, Huberman M, Johnson J, Telewski FW (1985): Thigmomorphogenesis: The induction of callose formation and ethylene evolution by mechanical peturbation in bean stems. Physiol Planta 64:271–279.
Jones RS, Mitchell CA (1989): Calcium ion involvement in growth inhibition of mechanically stressed soybean (*Glycine max*) seedlings. Physiol Planta 76:598–602.
Julian FJ, Goldman DE (1962): The effects of mechanical stimulation on some electrical properties of axons. J Gen Physiol 46:297–313.
Kauss H (1987): Some aspects of calcium-dependent regulation in plant metabolism. Ann Rev Plant Physiol 38:47–72.
Kink JA, Maley ME, Preston RR, Ling K-Y, Wallen-Friedman MA, Saimi Y, Kung C (1990): Mutations in *Paramecium* calmodulin indicate functional differences between the C-terminal and N-terminal lobes *in vivo*. Cell 62:165–174.
Knight MR, Campbell AK, Smith SM, Trewavas AJ (1991): Transgenic plant acquorin reports the effects of touch and cold-shock and elicitors on cytoplasmic calcium. Nature 352:524–526.
Kung C (1971a): Genic mutations with altered system of excitation in *Paramecium aurelia*. I. Phenotypes of the behavioral mutants. Z Vergl Physiol 71:142–164.
Kung C (1971b): Genic mutations with altered system of excitation in *Paramecium aurelia*. II. Mutagenesis, screening and genetic analysis of the mutations. Genetics 69:29–45.
Lee VD, Huang B (1990): Caltractin: A basal body-associated calcium-binding protein in *Chlamydomonas*. In O'Day DH (ed): "Calcium as an Intracellular Messenger in Eukaryotic Microbes." Washington, DC: American Society for Microbiology, pp 245–257.
Ling V, Perera I, Zielinski RE (1991): Primary structures of *Arabidopsis* calmodulin isoforms deduced from the sequences of cDNA clones. Plant Physiol 96:1196–1202.
Lukas TJ, Wallen-Friedman M, Kung C, Watterson DM (1989): *In vivo* mutations of calmodulin: a mutant *Paramecium* with altered ion current regulation has an isoleucine-to-threonine change at residue 136 and an altered methylation state at lysine residue 115. Proc Natl Acad Sci USA 86:7331–7335.

Martinac B, Buechner M, Delcour AH, Adler J, Kung C (1987): Pressure-sensitive ion channel in *Escherichia coli*. Proc Natl Acad Sci USA 84: 2297–2301.

Means AR (1988): Molecular mechanisms of action of calmodulin. Rec Prog Hormone Res 44:223–259.

Mitchell CA, Severson CJ, Wott JA, Hammer PA (1975): Seismomorphogenic regulation of plant growth. Amer Soc Hort Sci 100:161–165.

Morgan JM (1984): Osmoregulation and water stress in higher plants. Ann Rev Plant Physiol 35:299–319.

Okazaki Y, Tazawa M (1986): Involvement of calcium ion in turgor regulation upon hypotonic treatment in *Lamprothamnium succintum*. Plant Cell Environ 9:185–190.

Okazaki Y, Yoshimoto Y, Hiramoto Y, Tazawa M (1987): Turgor regulation and cytoplasmic free Ca^{2+} in the alga *Lamprothamnium*. Protoplasma 140:67–71.

Pickard BG (1971): Action potentials resulting from mechanical stimulation of pea epicotyls. Planta 97:106–115.

Pickard BG (1984): Voltage transients elicited by a sudden step-up of auxin. Plant Cell Environ 7:679–682.

Roux SJ, Slocum RD (1982): Role of calcium in mediating cellular functions important for growth and development in higher plants. In Cheung WY (ed): "Calcium and Cell Function, Vol III." New York: Academic Press, pp 409–453.

Sachs F (1988): Mechanical transduction in biological systems. Critical Rev Biomed Eng 16:141–169.

Saimi Y, Kung C (1980): A Ca^{2+}-induced Na^{+} current in *Paramecium*. J Exp Biol 88:305–325.

Saimi Y, Martinac B, Gustin MC, Culbertson MR, Adler J, Kung C (1988): Ion channels in *Paramecium*, Yeast and *Escherichia coli*. Trends Biochem Sci 13:304–309.

Schaefer WH, Hinrichsen RD, Burgess-Cassler A, Kung C, Blair LA, Watterson DM (1987): A mutant *Paramecium* with a defective calcium-dependent potassium conductance has an altered calmodulin: a nonlethal selective alteration in calmodulin regulation. Proc Natl Sci USA 84:3931–3935.

Sibaoka T (1969): Physiology of rapid movements in higher plants. Ann Rev Plant Physiol 20:165–184

Sinnott EW (1960): "Plant Morphogenesis." New York: McGraw-Hill, pp 43–54.

Stoclet J, Gerard D, Kilhoffer M, Lugnier C, Miller R, Schaeffer P (1987): Calmodulin and its role in intracellular calcium regulation. Prog Neurobiol 29:321–364.

Vantard M, Lambert A-M DeMey J, Picquot P, VanEldik LJ (1985): Characterization and immunocytochemical distribution of calmodulin in higher plant endosperm cells: Localization in the mitotic apparatus. J Cell Biol 101:488–499.

Wick SM, Muto S, Duniec J (1985): Double immunofluorescense labeling of calmodulin and tubulin in dividing plant cells. Protoplasma 126:198–206.

Wirtz HRW, Dobbs LG (1990): Calcium mobilization and exocytosis after one mechanical stretch of lung epithelial cells. Science 250:1266–1269.

Yang XC, Sachs F (1989): Block of stretch-activated ion channels in *Xenopus* oocytes by gadolinium and calcium ions. Science 243:1068–1071.

Evolutionary Conservation of Developmental Mechanisms, pages 199–212

13. Evolutionary Conservation of Developmental Mechanisms: The *Neurospora* Mating Type Locus

Charles A. Staben

School of Biological Sciences, University of Kentucky,
Lexington, Kentucky 40506

INTRODUCTION

The theme of this symposium recognizes that the life cycles of different organisms are related to one another by evolutionary history. Underlying similarities of development in vastly different organisms suggests that diversity has arisen by rearrangement and alteration of basic biological building blocks. Developmental regulator genes, like the mating type genes of fungi and effector genes which are their targets, are likely to be important building blocks. To understand development, evolution, and the relation between them, we must understand how these genes act and how they have evolved.

The mating type genes of *Neurospora crassa* have multiple roles in regulating the *Neurospora* life cycle. Their crucial roles in sexual reproduction also make them important regulators of their own transmission. I will describe the roles that mating type plays in regulating the *Neurospora* life cycle, our knowledge of the mating type genes, and propose a model to explain the action of the *a* mating type DNA. In addition, I will describe the relationship of the *a-1* gene product to other regulatory proteins and, finally, I will compare mating type in *N. crassa* to similar systems in closely related fungi with different life cycles.

NEUROSPORA CRASSA LIFE CYCLE AND ROLES OF MATING TYPE

N. crassa is a haploid, heterothallic filamentous fungus. The fungus differentiates specialized asexual and sexual structures during the life cycle. Mating type has roles during vegetative growth and sexual development. The life cycle of *N. crassa* is diagrammed in Figure 1. Figure 2 contains micrographs of various developmental stages.

The most prominent vegetative form of *N. crassa* is the multinucleate mycelium (Fig. 2a). Two types of asexual spores, macroconidia and microconidia, develop from mycelium under appropriate environmental

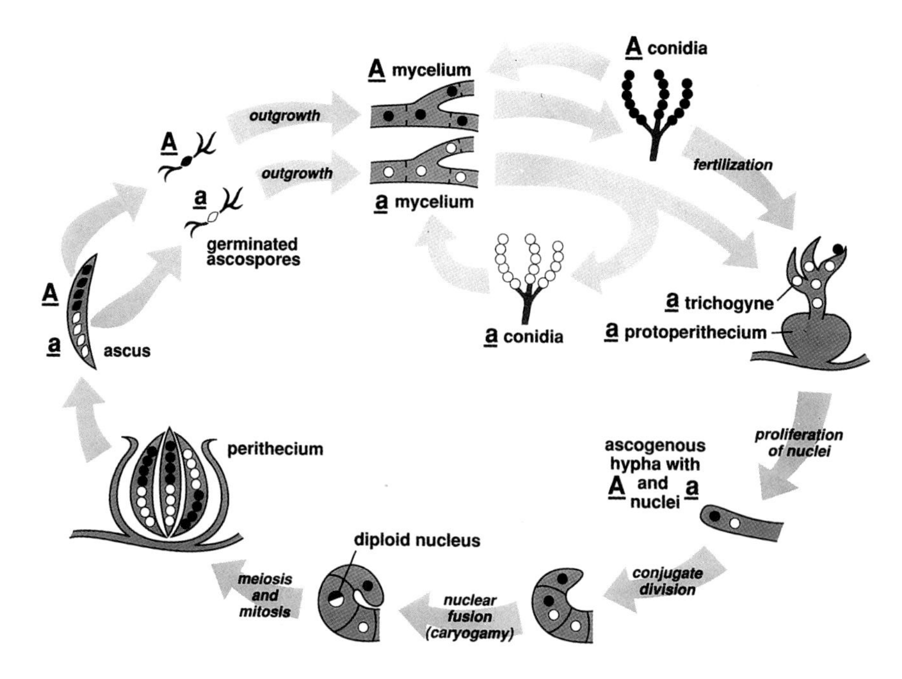

Fig. 1. Life cycle of *Neurospora crassa*.

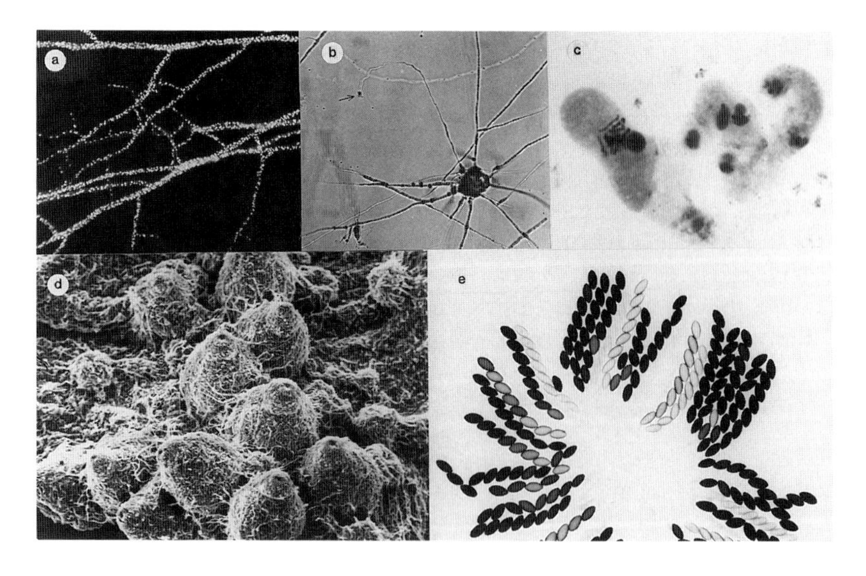

Fig. 2. *Neurospora crassa* development. The following micrographs depict different stages in the development of *N. crassa*. **a**: Fluorescence micrograph of multinucleate vegetative mycelium stained with a DNA-binding dye (Hoechst 33258). **b**: A trichogyne emanating from a protoperithecium (lower right) has grown toward a conidium of opposite mating type (arrow). **c**: (Right) Crozier at last premeiotic mitosis. (Center) A three-celled crozier before karyogamy. (Left) Condensed chromosomes during meiosis. **d**: Scanning electron micrograph of perithecia on agar surface. **e**: A rosette of maturing asci, each ascus containing eight ascospores. Figure reprinted from (26). (Micrographs adapted from: A. Springer, M.L. and Yanofsky, C. (1989) Genes Dev. 3:559–571; b. Bistis, G.N. (1981) Mycologia 73:967; c,e from Raju, N.B. (1980) 23:210,219; d. provided by M.L. Springer.)

conditions. Mating type has two roles during vegetative growth. First, mating type acts as a *het*erokaryon incompatibility locus (1). This function prohibits the formation of viable *a*+*A* heterokaryons during vegetative growth. Mating type is one of ten identified *het* loci (2). The vegetative incompatibility due to mating type, but not that due to other *het* loci, is relieved by mutation at the *tol*(erant) locus (3). Therefore, the heterokaryon incompatibility due to mating type is thought to require the *tol* gene product, and be partly (or wholly) independent of other *het* genes. Second, mating type determines male pheromone production. Any vegetative form can act as a male during sexual reproduction. Males produce mating-type-specific pheromones that orient the growth of female hyphae towards them (Fig. 2b), mediating fertilization (4). These hormones are not produced in mating type mutants that are infertile (4).

Sexual determination in *Neurospora*, unlike that described in *Drosophila melanogaster, Caenorhabditis elegans,* or the mouse, is largely environmental rather than genetic, and females are the only sexually differentiated form. Nitrogen starvation induces either mating type to develop female structures: protoperithecia and trichogynes. Fusion of a trichogyne with a male of opposite mating type initiates sexual reproduction, which culminates in meiosis and the production of sexual progeny. Sexual development is concomitant with complex morphological development of the female fruiting body, the perithecium (Fig. 2d). Mating occurs between males and differentiated females only of opposite mating type. The mating type genes control several individual steps within the sexual reproduction pathway. First, males produce mating-type-specific pheromones. These pheromones orient the growth of female hyphae of opposite mating type towards the male (4), a mating-type-specific response. After the female hypha fuses with the male, a single male nucleus enters the incipient fruiting body (protoperithecium). Fusion of opposite mating types induces morphological development and the fertilized perithecia enlarge and darken. Within the perithecium, the *a* and *A* nuclei divide several times. Eventually, one nucleus of each mating type is partitioned into special dikaryotic cells. Dikaryons immediately undergo karyogamy and meiosis to form eight-spored asci (Fig. 2c,e) that always contain four *a* and four *A* spores. A mating-type-dependent mechanism must assure that one *a* and one *A* nucleus are partitioned into dikaryons that produce ascospores. Either nuclei are identified by mating type, or only dikaryons containing one *a* and one *A* nucleus develop into asci.

THE *NEUROSPORA* MATING TYPE LOCUS

The mating type of an *N. crassa* strain is determined solely by the DNA sequence present at its mating type locus. The *a* mating type is encoded in 3235 bp of DNA (5) unique to *a* strains; *A* mating type is encoded by 5301 bp of DNA unique to *A* strains (6). The structure of the locus is diagrammed in Figure 3. Unique DNA segments that occupy the same locus, but that may contain multiple genes, are termed idiomorphs (7). The idiomorphic DNA at the mating type locus is both necessary and sufficient to determine mating type. The results of thousands of carefully controlled *Neurospora* crosses suggested that otherwise isogenic strains will mate provided only that they are of opposite mating type. We have shown that strains of opposite mating type contain different idiomorphic DNAs (5,6). In fact, substitution of one sequence for the other by direct gene replacement changes mating type, allowing otherwise isogenic strains to mate (Chang and C.S., unpublished observations). *N. crassa*, unlike *S. cerevisiae*, contains no silent mating type cassettes (nor does *N. crassa* switch mating type).

FUNCTIONAL ANALYSIS OF THE MATING TYPE LOCUS

Transformation assays indicate that the *a* idiomorph (5) contains at least two functional regions. One region, *a-1*, encodes mating identity, perithecium induction, and vegetative incompatibility. The second region, *a-2*, is required in addition to *a-1* for perithecial maturation of dual mating type transformants. The *A* idiomorph also appears to contain at least two distinct functional regions (8).

Functions of the *a-1* Region

Both transformation assays and characterization of mating mutants indicate that *a-1* has essential roles in mediating vegetative incompatibility, determining the mating identity of males and females, and inducing perithecium development. This DNA is not sufficient to confer fertility on mating

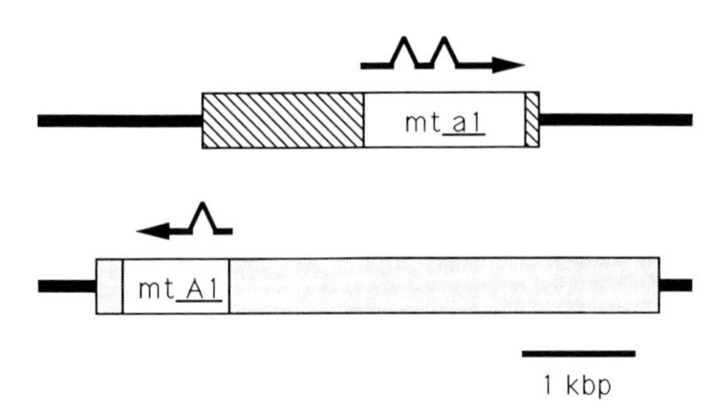

Fig. 3. Structure of the mating type locus. Unique regions of the mating type idiomorphs, represented by different shading patterns, are bordered by identical genomic sequences, solid black. Presumed transcripts are indicated by the thin arrows, with introns indicated by indentations. The *a-2* region of *a* is within the diagonally shaded box to the left of mt *a-1*, the *A-2* region within the stippled area adjacent to mt *A-1*.

mutants or self-fertility on *A* strains. The *a-1* region is defined as the smallest segment of DNA capable of conferring *a* properties upon appropriate recipients. It appears that the function of this DNA is to produce a polypeptide, mt *a-1*, described in Section V.

DNA introduced into the *N. crassa* genome by transformation typically integrates nonhomologously at random sites. When *a* idiomorph DNA is transformed into an *a* mutant, such as a^{m1} (a strain that has no mating type vegetative incompatibility and produces no pheromone or any mating response), both vegetative incompatibility and mating properties are restored. The *a* DNA can also be transferred to a *tol, A* strain, in which case the strain becomes able to mate as both *a* and *A* (described more fully below). A small portion of the idiomorphic DNA, from nucleotides 2923–4596 of the published sequence, confers all *a* properties (Fig. 3).

Analysis of mutants derived from *a* strains impaired in mating type function also indicates that *a-1* is an important functional region. Mutants a^{m1}, a^{m30} (nulls), and a^{m33} (mates, vegetatively compatible) all contain altered *a-1* DNA. Cloned DNAs bearing these alterations have the expected properties in transformation assays; repair of the defects in the cloned DNAs restores full activity (5). Additional details of these mutants are discussed below in Structure and Functions of the mt *a-1* Polypeptide.

Functions of the *a-2* Region

The *a-2* region is idiomorphic DNA, aside from *a-1*, that confers properties additional to those conferred by *a-1* upon transformants (5). Transformants bearing this additional 1.4 kbp of idiomorphic DNA in *cis* with *a-1* have two additional properties. When *tol, A* transformants bearing only *a-1* are mated to *A* strains, perithecia are induced, but differentiation of the perithecia ceases at an early developmental stage. Transformants bearing the entire idiomorph differentiate well-developed perithecia, though they are not fertile. Apparent self-fertilizations of the two types of transformants also differentiate to a different extent. The idiomorphic DNA in this region confers no apparent phenotype upon transformants that bear it alone. No mRNA product from the region has been detected in liquid-grown cultures. Finally, open reading frames within the region do not match *Neurospora* codon usage patterns nor are they homologous to protein sequences encoded by the DNA data bases or deposited in protein data bases. The *a-2* region may encode a novel polypeptide, produced only within specialized cells inside the developing perithecium. Alternatively, this region may act as a DNA sequence, perhaps to regulate the activity of the *a-1* region. The role of this region in normal ascosporogenesis has not been established. Transformants in which residues 1665–2127 have been replaced by a selectable marker differentiate normally, so the activity apparently does not depend upon this segment.

Action of Mating type DNA in Dual-Mating Type Transformants

Strains that bear two different idiomorphs are essentially infertile. Strains of *tol, A* transformed with *a* idiomorph DNA will mate and form perithecia when crossed with either *A* or *a* strains. Conversely, *tol,a* strains transformed with *A* idiomorph DNA will mate and form perithecia when crossed with either *a* or *A* strains. However, these perithecia are typically barren. A small number of viable spores are produced from rare perithecia. (These perithecia could be the result of fertilizations by heterokaryotic transformants in which single-mating type nuclei are still present.) We expect independent assortment of the resident- and transforming-mating type DNAs would result in the formation of dual-mating type progeny from crosses fertilized with a dual-mating parent. However, of over 200 spores from such crosses, none has had a dual-mating type. Additionally, all dual-mating transformants are self-sterile, rather than self-fertile (Chang and C.S., unpublished observations). These observations suggest that karyogamy between nuclei in which one partner bears both mating type genes is unproductive or cannot occur. Regulation of karyogamy by mating type is consistent with the observations of many *Neurospora* geneticists, derived from thousands of crosses, that every ascus contains 4 *a* and 4 *A* spores.

Developmental Timing of Mating Type Gene Action

Transcripts of the *a-1* gene are present in *a* strains in vegetatively growing liquid cultures. The levels of mRNA are extremely low (5), and have not been assayed throughout the life cycle. A functional *a-1* gene is necessary for pheromone induction or mating, but it is not necessary for differentiation of females, (though females lacking the function are unresponsive to pheromone and infertile).

An unusual genome modification process in *N. crassa* offers a unique means of assessing the stage-specific action of DNA sequences. Repeat induced point mutation (RIP) is a novel process that introduces point mutations throughout all copies of duplicated sequences before meiosis in *N. crassa* (9,10). This process has been used to disrupt mating type genes. Transformants that bear duplications of the entire *a* locus are fertile, though they generate infertile progeny. These progeny have not been completely analyzed, but the infertility is presumably due to mutation of both copies of the *a* idiomorph DNA by RIP (C.S., unpublished observations). This experiment suggests that intact *a-1* and *a-2* regions are not essential for steps in ascosporogenesis that occur after RIP. (The products of *a-1* or *a-2* could be stored prior to inactivation and used after inactivation, but clearly the sequences themselves are not necessary for steps after RIP.) Similar experiments support similar conclusions with respect to the *A* idiomorph (8).

STRUCTURE AND FUNCTIONS OF THE mt *a-1* POLYPEPTIDE

The *a-1* region encodes a 382-amino-acid polypeptide (mt *a-1*) that mediates vegetative incompatibility and mating. Mt *a-1* is encoded by a spliced mRNA transcribed at low levels in liquid media, either nitrogen-rich or nitrogen-limited (vegetative or mating conditions). Strains selected for escape from mating-type vegetative incompatibility (11) have mutations in *a-1* responsible for their mating type defects (5,11). Two such mutants have lost both vegetative incompatibility and mating type activities, and contain a deletion (a^{m1}) or insertion (a^{m30}) within *a-1*. Another mutant, a^{m33}, can mate but has no vegetative incompatibility function. This mutant has a single base change in *a-1* that changes residue 258 of mt *a-1* from Arg to Ser.

The mt *a-1* HMG Box

The structure of mt *a-1* and its similarity to other proteins suggest that it is a DNA binding protein. The *a-1* polypeptide sequence is similar to the *S. pombe* mat-M_c (12) and *Podospora anserina* Mat-P1 (M. Picard and E. Coppin, personal communication) polypeptides. The similarities in sequences and mating function suggest that the polypeptides are homologous. These polypeptides contain an essentially identical 12-amino-acid segment that presumably plays a critical role in mating. Mt *a-1*, mat-M_c, and mat-P1 are also similar to a subset of high mobility group (HMG) proteins and other regulatory gene products—transcription activators (13–17), a transcription inhibitor (18), and male-determining factors (SRY proteins) of unknown function (19,20). The similar portions of these sequences are aligned in Figure 4. The HMG proteins are nonhistone, chromatin-associated proteins that bind to DNA (21). HMG-1 and HMG-2 proteins appear to be class II

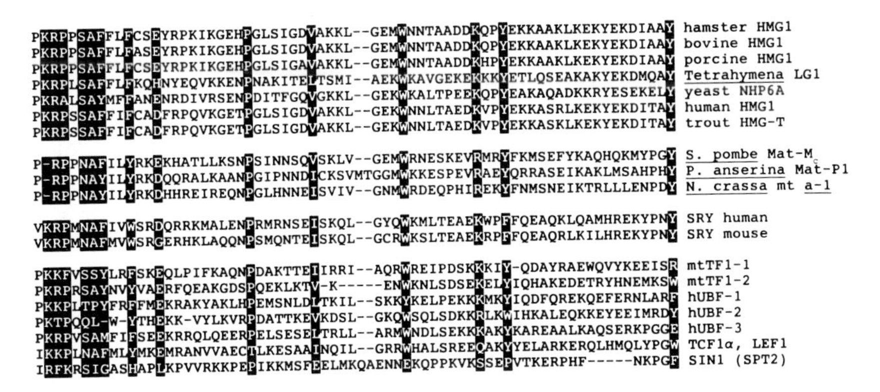

Fig. 4. Alignment of HMG box proteins. The HMG box portions of polypeptides described in the text are aligned to show conservation of sequence. Residues conserved or conservatively substituted in at least 15 of 18 sequences are shown in reverse print. Origin of each sequence is at right.

transcription factors (22). The similarity of the mating polypeptides to these other proteins is restricted to the region called the HMG box that is required for DNA binding (13,14). This HMG box region is the segment of near-perfect homology between mt *a-1*, mat-P1, and mat-M_c. The presence of an HMG box strongly suggests that the *N. crassa* mt *a-1, P. anserina* mat-P1, and *S. pombe* mat-M_c polypeptides are DNA-binding proteins.

Multiple Functional Domains in Mt *a-1*

HMG box proteins have many different functions in gene regulation. The function may depend upon the precise sequence of the HMG box, upon the presence of different functional domains within the polypeptide, or upon the association of the HMG box protein with another polypeptide (23). Analysis of mt *a-1* suggests that it has at least two functional domains, one of which contains the HMG box.

The first 200 amino acids of mt *a-1* probably form a DNA-binding domain essential for mating type activity, while the carboxy-terminal half is required for vegetative incompatibility and may be a transcription activation domain. The amino-terminal 200 amino acids of mt *a-1* are homologous to the *S. pombe* Mat-Mc and the *P anserina* Mat-P polypeptides. This region of mt *a-1*, and its homologs in that Mat-Mc and Mat-P1, contain HMG boxes and are very basic, as expected for DNA-binding domains. Two observations suggest that the vegetative incompatibility function of mt *a-1* depends upon a C-terminal domain. First, neither *S. pombe* nor *P. anserina* have mating-type-associated vegetative incompatibility systems. The *S. pombe* mat-Mc protein completely lacks amino acids corresponding to the last 180 amino acids of mt *a-1*. The similarity of *P. anserina* mat-P to mt *a-1* is only 12% in the last 150 amino acids, much less than the 31% identity in the amino-terminal 200 amino acids. Second, the a^{m33} mutation, which specifically inactivates the vegetative incompatibility function, is a change of residue 258 from Arg to Ser (5). The sequence of the C-terminal half of mt *a-1*, like the activation domains of many transcriptional activators, is acidic and proline-rich. Within the carboxyl terminus of mt *a-1*, 18 of the final 100 amino acids are proline, and this segment, in contrast to the amino-terminal domain, is acidic. (Although the sequences of these regions of mt *a-1* and Mat-P1 are different, both mt *a-1* and Mat-P1 sequences are acidic and proline-rich. The presence of similar acidic domains in mt *a-1* and mat-P1 and the absence of this region in mat-M_c suggests either that the first two have additional or different functions not found in the *S. pombe* polypeptide, or the mat-M_c combines with another polypeptide that provides these functions.) Proteins that activate transcription frequently contain acidic domains (24). For example, the transcriptional activation domain of human CTF/NF-1 is a 100-amino-acid segment that contains 23 prolines (25). The HMG box containing T-cell transcription factor also contains a proline-rich domain (14,15,17). These observations suggest that mt *a-1* binds to DNA and acts as a transcription factor.

ORGANIZATION AND FUNCTIONS OF THE A IDIOMORPH

Though dissimilar in sequence, the *A* and *a* idiomorphs resemble one another in organization (5,6,26). Like *a, A* has at least two functional regions. Mt *A-1* encodes vegetative incompatibility and mating functions. Both activities reside in a 288 amino acid polypeptide, mt *A-1*, produced from a spliced mRNA. Frameshift mutations (6) within the mt *A-1* coding region eliminate both *A* activities. Mt *A-1* is homologous to the *S. cerevisiae* α1 mating type gene (27). The α1 polypeptide combines with another polypeptide, PRTF (MCM1), to activate α-specific genes (28,29). The other portion of *A (A-2)* is required for development of fertile perithecia. However, *A-2* can be inactivated by RIP (8). Therefore, *A-2* DNA must not be required after RIP. No *A-2* product has been identified, nor has the *A-2* region been precisely mapped.

MODEL FOR MATING TYPE CONTROL OF THE LIFE CYCLE

My model for mating type control of the life cycle, shown in Figure 5, (26) is analogous to that by which the *S. cerevisiae* MAT locus is thought to regulate the yeast life cycle. Genes at the yeast MAT locus encode regulatory proteins that determine the specialized properties of each haploid mating type by causing mating type-specific gene expression. Fusion between cells of opposite mating type generates a new cell type, the diploid, in which products from the mating type genes form novel regulatory species that activate sets of target genes distinct from those active in either haploid (30). Such a model must account for the roles of the mating type genes during vegetative growth and during each step in sexual development.

During vegetative growth, *a-1* is expressed (5) and causes expression of *a* specific genes, including those controlling expression of *a* pheromone. Mutants in *a-1* do not secrete active *a* pheromone (4). *A-1* plays a similar role in *A* strains. If *a* and *A* mycelia fuse, stable heterokaryon formation is prevented unless *a-1, A-1*, or *tol* has been mutated. The vegetative incompatibility reaction is apparently due to the interaction of mt *a-1* and mt *A-1* (or factors controlled by them), which are now present in the same cytoplasm. This interaction is dependent on other genes, including *tol* (3).

The *a-1* gene is also expressed under nitrogen-limiting conditions that induce sexual differentiation (5). Because *a-1* is required for female responsiveness to *A* pheromone, we know that *a-1* must induce a mating type specific pheromone receptor or response pathway in females (4). The *a* idiomorph is also required for morphological development and ascosporogenesis, as described earlier. I propose that the interaction of the *a*

Vegetative

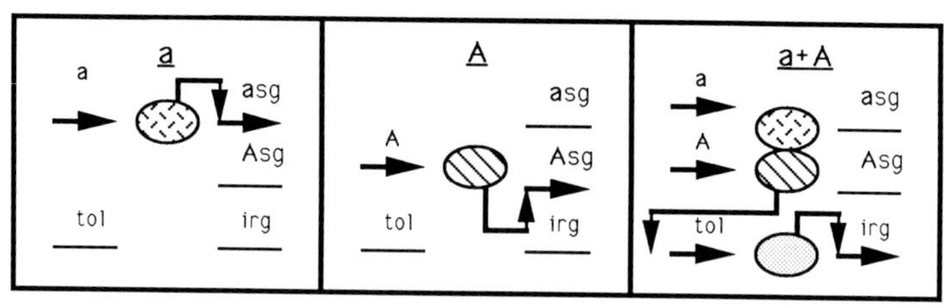

Sexual

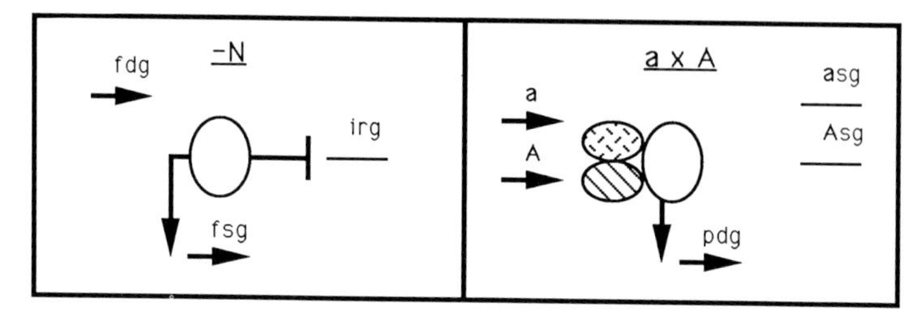

Fig. 5. Model for mating type gene action. Under vegetative conditions (**top panel**) *a* induces expression of *a*-specific genes; *A* induces expression of *A*-specific genes. Fusion of *a* and *A* induces an incompatibility response (induction of *i*ncompatibility *r*esponse *g*enes) dependent upon the *tol* gene. In the sexual cycle, nitrogen limitation induces female differentiation genes (fdg) that induce female-specific genes (fsg). The incompatibility response is inactivated in females, and the interaction of *a* and *A* cells results in formation of a novel regulator that induces expression of genes necessary for perithecium differentiation (pdg). This model represents one of many possible genetic control systems; many of the proposed regulatory elements and their interactions have not been demonstrated. Figure adapted from (26).

and *A* idiomorphs, or their products after fertilization, is required for the gene expression necessary for sexual development. The result of this interaction is different from that which occurs during vegetative growth, so either the mating type gene products or factors with which they interact are modified during female differentiation. Our transformation results (5) indicate that *a-2* has a role in development of perithecia. The gene disruption experiments of Glass (8) indicate that the *A-2* region is required for events after mating that are required for formation of fertile perithecia. These data suggest that the *a-2* and *A-2* regions or products encoded by these regions act in the croziers to identify zygotes formed from *a* and *A* nuclei. Two simple models for *a-2* function suggest that the *a-2* region may act as a DNA sequence, to regulate, for example, expression of *a-1*, or that it may encode a diffusible product. Such a product may be expressed only in dikaryotic or zygotic tissue. The apparent sterility of dual-mating type transformants suggests that the mechanism monitoring or promoting successful crozier formation is sensitive to the mating type of the fusing nuclei.

Undoubtedly, *Neurospora* mating type genes regulate target genes. Different sets of target genes may be important during vegetative growth and sexual development. To understand mating, these targets must be identified, and their interaction with the mating type genes characterized. Only one gene that depends upon *a/A* interaction, *tol* (3) has been identified.

EVOLUTIONARY CONSIDERATION OF MATING TYPE CONTROLS

A sequence relationship of the SRY protein, a sex determination factor in mammals, to *N. crassa* mt *a-1* polypeptide encourages speculation about a fundamental relationship between sex control mechanisms. However, the similarity between SRY and mt *a-1* is confined to the HMG box and is no greater, and probably no more significant, than the relationship of either polypeptide to other HMG box polypeptides. The presence of conserved sequence motifs, such as the HMG box motif, in diverse regulatory proteins is a familiar theme in molecular biology.

Sexual reproduction is a potentially dangerous game to play with one's genes, and in most organisms the process is carefully regulated. In fungi, sexual reproduction can be regulated at each of the fundamental steps: sexual differentiation, plasmogamy, karyogamy, and meiosis. Regulation of these steps produces diverse life cycles. One would like to determine whether similar regulators play different roles in organisms with diverse life cycles, and how these life cycles may have arisen by modification of regulatory factor action during evolution.

For example, the *S.pombe* mat-M_c and *N. crassa* mt *a-1* polypeptides are apparently homologs and are presumed to play similar biochemical roles. However, the actions of these polypeptides in the context of the respective life cycles are quite different. In *S. pombe*, but not in *N. crassa*, mating occurs without obvious sexual differentiation. Also, the *S. pombe* diploids formed by mating can propagate vegetatively, unlike *N. crassa* diploids, which immediately undergo meiosis. The evolution of these diverse life cycles probably depended on changes in the mating factors, in the regulation of mating factor expression, and in the consequences of mating type gene action.

In addition to comparing the sequence of mating type polypeptides, we have examined close relatives of *N. crassa* for the presence of DNA sequences related to the *a* and *A* idiomorphs. All members of the genus *Neurospora* with distinct mating types have either an *a* or an *A* DNA homolog (31). One member of the genus, *Neurospora tetrasperma*, propagates as a self-fertile, vegetatively compatible *a/A* heterokaryon. Transfer of its mating type genes to *N. crassa* yields a fertile hybrid with mating type vegetative incompatibility (32). Therefore, the different life cycle exhibited by *N. tetrasperma* must involve a change in the target genes that mediate vegetative incompatibility rather than in the mating type genes themselves.

Six strains within the genus *Neurospora* have a homothallic life cycle; haploids propagate and are self-fertile. One strain has both *A* and *a* homologous sequences. Five homothallic isolates have *A* homologs, but no apparent *a* homolog (31). One of the latter strains bears a *A* idiomorph that functions in *both* mating and vegetative incompatibility in *N. crassa* (7), though the mating type idiomorphs must not have an incompatibility function in their native species. These observations pose several puzzling questions about mating type and the evolution of sexual life cycles within the genus *Neurospora*. First, do homothallic isolates without an *a* homolog contain an alternate mating type idiomorph, invisible to the *a* DNA probes because of sequence diversity, that serves the *a* function? Second, in view of the reduced fertility of dual-mating type *N. crassa* strains and our inability to transmit both *a* and *A* idiomorphs to meiotic progeny, how do strains that have copies of both idiomorphs avoid infertility?

Questions about mating strategies may be clarified by consideration of the phylogenetic context of mating in the fungi. Within the *Ascomycete* lineage, heterothallic species with two mating types are common. Homothallic life cycles appear in many lineages, and the mechanisms accounting for homothallism appear to be diverse. A parsimonious explanation is that heterothallic mating is ancestral in *Ascomycetes* and that it must have arisen before, or soon after, the divergence of the *Ascomycete* lineage from that of other fungi (assuming that the order is monophyletic). Variations in life cycle have occurred frequently. Mutations such as a^{m33} prove that the vegetative incompatibility and mating functions can be separated. The retention of vegetative incompatibility in mating type genes of species within the genus *Neurospora* where there is no obvious role for this function in the native species, suggests that variation in life cycle frequently results from variation in the action of mating type target genes that is more rapid than evolutionary change in the mating type genes themselves.

ACKNOWLEDGMENTS

Much of the work described here was done as a postdoctoral researcher with Charles Yanofsky at Stanford University. Work at the University of Kentucky has been supported by the University of Kentucky, by Grant IN163 from the American Cancer Society, and by BRSG-SO7-RR07114-21 awarded by the National Institutes of Health. I would also like to acknowledge Chang Shinyu and Melissa Philley for their participation in recent research and in the preparation of this manuscript.

REFERENCES

1. Beadle, GW, Coonradt VL 1944: Heterocaryosis in *Neurospora crassa*. Genetics **29**:291–308.
2. Perkins DD, Radford A, Newmeyer D, Bjorkman M 1982: Chromosomal loci of *Neurospora crassa*. Microbiol Rev **46**:462–570.

3. Newmeyer D 1970: A Suppressor of the heterokaryon-incompatibility associated with mating type in *Neurospora crassa*. Can J Gen Cytol **12**:914–926.
4. Bistis GN 1983: Evidence for diffusible mating-type specific trichogyne attractants in *Neurospora crassa*. Exp Mycol **7**:292–295.
5. Staben C, Yanofsky 1990: *Neurospora crassa* A mating type region. Proc Natl Acad Sci USA **87**:4917–4921.
6. Glass NL, Grotelueschen J, Metzenberg RL 1990 *Neurospora crassa* A mating-type region. Proc Natl Acad Sci USA **87**:4912–4916.
7. Metzenberg RL, Glass NL 1990: Mating type and mating strategies in *Neurospora*. Bioessays **12**:53–59.
8. Glass NL 1991: Mating type in *Neurospora*. Fun Genet Newsl **38**:10.
9. Selker EU, Cambareri EB, Jensen BC, Haack KR 1987: Rearrangement of duplicated DNA in specialized cells of Neurospora. Cell **51**:741–752.
10. Cambareri EB, Jensen BC, Schabtach E, Selker EU 1989: Repeat-induced G-C to A-T mutations in *Neurospora*. Science **244**: 1571–1575.
11. Griffiths AJF, DeLange AM 1978: Mutations of the a mating-type gene in *Neurospora crassa*. Genetics **88**: 239–254.
12. Kelly M, Burke J, Smith M, Klar AJS, Beach D 1988: Four mating-type genes control sexual differentiation in the fission yeast. EMBO J **7**:1537–1547.
13. Jantzen H-M, Admon A, Bell SP, Tjian R 1990: Nucleolar transcription factor hUBF contains a DNA-binding motif with homology to HMG proteins. Nature **344**:830–836.
14. Waterman ML, Fischer WH, Jones KA 1991: A thymus-specific member of the HMG protein family regulates the human T cell receptor Cα enhancer. Genes Dev **5**:656–669.
15. Travis A, Amsterdam A, Belanger C, Grosschedl R 1991: LEF-1, a gene encoding a lymphoid-specific with protein, an HMG domain, regulates T-cell receptor α enhancer function. Genes Dev **5**:880–894.
16. Parisi MA, Clayton DA 1991: Similarity of human mitochondrial transcription factor 1 to high mobility group proteins. Science **252**: 965–969.
17. van de Wetering M, Oosterwegel M, Dooijes D, Clevers H 1991: Identification and cloning of TCF-1, a T-lymphocyte-specific transcription factor containing a sequence-specific HMG box. EMBO J **10**:123–132.
18. Roeder GS, Beard C, Smith M, Keranen S 1985: Isolation and characterization of the SPT2 gene, a negative regulator of Ty-controlled yeast gene expression. Mol Cell Biol **5**: 1543–1553.
19. Sinclair AH, Berta P, Palmer MS, Hawkins JR, Griffiths BL, Smith MJ, Foster JW, Frischauf A-M, Lovell-Badge R, Goodfellow PN 1990: A gene from the human sex-determining region encodes a protein with homology to a conserved DNA-binding motif. Nature **346**: 240–244.
20. Gubbay J, Collignon J, Koopman P, Capel B, Economou A, Munsterberg A, Vivian N, Goodfellow PN, Lovell-Badge R 1990: A gene mapping to the sex-determining region of the mouse Y chromosome is a member of a novel family of embryonically expressed genes. Nature **346**:245–250.
21. Wright JM, Dixon GH 1988: Induction by torsional stress of an altered DNA conformation 5′ upstream of the gene for a high mobility group protein from trout and specific binding to flanking sequences by the gene product HMG-T. Biochem **27**:576–581.
22. Singh J, Dixon GH 1990: High mobility group proteins 1 and 2 function as general class II transcription factors. Biochem **29**:6295–6302.
23. Bell SP, Jantzen, H-M, Tjian R 1990: Assembly of alternative multiprotein complexes directs rRNA promoter selectivity. Genes Dev **4**:943–954.
24. Ptashne M 1988: How eukaryotic transcriptional activators work. *Nature* **335**:683–689.
25. Mermod N, O'Neill EA, Kelly TJ, Tjian R 1989: The proline-rich transcriptional activator of CTF/NF-1 is distinct from the replication and DNA binding domain cell **58**: 741–753.
26. Glass NL, Staben C 1990: Genetic control of mating in *Neurospora crassa* Sem Dev Biol **1**: 177–184.

27. Astell C, Ahlstrom-Jonasson L, Smith M, Tatchell K, Nasmyth KA, Hall BD 1981: The sequence of DNAs coding for the mating-type loci of *Saccharomyces cerevisiae*. Cell **27**:15–23.
28. Passmore S, Elble R, Tye B-K 1989: A protein involved in minichromosome maintenance in yeast binds a transcriptional enhancer conserved in eukaryotes. Genes Dev **3**:921–935.
29. Jarvis EE, Clark KL, Sprague GF 1989: The yeast transcription factor, PRTF, a homolog of the mammalian serum response factor, is encoded by the MCM1 gene. Genes Dev **3**:936–945.
30. Herskowitz I 1988: Life cycle of the budding yeast Saccharomyces cerevisiae. Microbiol Rev **52**: 536–553.
31. Glass NL, Vollmer SJ, Staben C, Metzenberg RL, Yanofsky C 1988: DNAs of the two mating-type alleles of *Neurospora crassa are highly dissimilar*. Science **241**: 570–573.
32. Metzenberg RL, Ahlgren K 1973: Behaviour of *Neurospora tetrasperma A* mating type genes introgressed into *N. crassa*. Can J Genet Cytol **15**:571–576.

Index